tredition

tredition was established in 2006 by Sandra Latusseck and Soenke Schulz. Based in Hamburg, Germany, tredition offers publishing solutions to authors and publishing houses, combined with worldwide distribution of printed and digital book content. tredition is uniquely positioned to enable authors and publishing houses to create books on their own terms and without conventional manufacturing risks.

For more information please visit: www.tredition.com

TREDITION CLASSICS

This book is part of the TREDITION CLASSICS series. The creators of this series are united by passion for literature and driven by the intention of making all public domain books available in printed format again - worldwide. Most TREDITION CLASSICS titles have been out of print and off the bookstore shelves for decades. At tredition we believe that a great book never goes out of style and that its value is eternal. Several mostly non-profit literature projects provide content to tredition. To support their good work, tredition donates a portion of the proceeds from each sold copy. As a reader of a TREDITION CLASSICS book, you support our mission to save many of the amazing works of world literature from oblivion. See all available books at www.tredition.com.

 Project Gutenberg

The content for this book has been graciously provided by Project Gutenberg. Project Gutenberg is a non-profit organization founded by Michael Hart in 1971 at the University of Illinois. The mission of Project Gutenberg is simple: To encourage the creation and distribution of eBooks. Project Gutenberg is the first and largest collection of public domain eBooks.

Man And His Ancestor A Study In Evolution

Charles Morris

Imprint

This book is part of TREDITION CLASSICS

Author: Charles Morris
Cover design: Buchgut, Berlin – Germany

Publisher: tredition GmbH, Hamburg - Germany
ISBN: 978-3-8472-1813-5

www.tredition.com
www.tredition.de

Copyright:
The content of this book is sourced from the public domain.

The intention of the TREDITION CLASSICS series is to make world literature in the public domain available in printed format. Literary enthusiasts and organizations, such as Project Gutenberg, worldwide have scanned and digitally edited the original texts. tredition has subsequently formatted and redesigned the content into a modern reading layout. Therefore, we cannot guarantee the exact reproduction of the original format of a particular historic edition. Please also note that no modifications have been made to the spelling, therefore it may differ from the orthography used today.

MAN AND HIS ANCESTOR

A STUDY IN EVOLUTION

BY

CHARLES MORRIS

AUTHOR OF "CIVILIZATION: AN HISTORICAL REVIEW OF ITS ELEMENTS," "THE ARYAN RACE," ETC.

New York

THE MACMILLAN COMPANY

LONDON: MACMILLAN AND CO., Ltd.

1900

All rights reserved

Copyright, 1900,

By THE MACMILLAN COMPANY.

Norwood Press
J. S. Cushing & Co. — Berwick & Smith
Norwood Mass. U.S.A.

PREFACE

It would be difficult to find any intelligent person in this age of the world who has not some theory or opinion in regard to the origin of man, and perhaps almost as difficult to find any such person who can give a good and sufficient reason for the faith that is in him. This is especially the case with those who look upon man as a product of evolution, a natural outgrowth from the world of lower life, since here simple faith or ancient authority is not sufficient, as in the creation hypothesis, but scientific evidence and logical argument are necessary. It is to enable this class of readers to test the quality and sufficiency of their belief that this book has been prepared.

The question of the evolutionary origin of man has been by no means neglected by recent authors, yet it has been dealt with chiefly as a side issue in works of a more extended purpose, and largely in technical language, simple to the scientist, but difficult to the general reader. The only work that makes this subject its leading theme, Darwin's "Descent of Man," adds to it a still longer treatise on "Sexual [vi] Selection," so that the subject of man's evolutionary origin cannot be said to have been yet dealt with for itself alone. Darwin's work, moreover, is now nearly thirty years old, and to this extent antiquated, while at best it cannot be considered as well suited for general reading.

These considerations have given rise to the present work, in which an effort has been made to present the subject of man's origin in a popular manner, to dwell on the various significant facts that have been discovered since Darwin's time, and to offer certain lines of evidence never before presented in this connection, and which seem to add much strength to the general argument.

The subject is one of such widespread interest as to make it probable that a plain and brief presentation of it will be acceptable, both to enable those who are evolutionists in principle to learn on what grounds their acceptance of this phase of evolution stands, and to aid those who are at sea on the whole subject of man's origin to reach some fixed conclusion. For these purposes this little book has been set afloat, with the hope that it may carry some doubters to solid land and teach some believers the fundamental elements of their faith.

MAN AND HIS ANCESTOR

I

EVOLUTION VERSUS CREATION

In any consideration of the origin of man we are necessarily restricted to two views: one, that he is the outcome of a development from the lower animals; the other, that he came into existence through direct creation. No third mode of origin can be conceived, and we may safely confine ourselves to a review of these two claims. They are the opposites of each other in every particular. The creation doctrine is as old almost as thinking man; the evolutionary doctrine belongs in effect to our own generation. The former is not open to evidence; the latter depends solely upon evidence. The former is based on authority; the latter on investigation. The doctrine of direct creation can merely be asserted, it cannot be argued; the statement once made, there is nothing more to be said; it is an *ipse dixit* pure and simple. The doctrine of evolution, on the contrary, founded as it must be on ascertained facts, is fully open to argument, and depends for its [2] acceptance on the strength and validity of the evidence in its favor.

If the doctrine of the direct creation of man had been originally presented in our own day, proof of the assertion would have been at once demanded, and the only evidence admissible would have been that of witnesses of the act of creation. There could, of course, have been no human witnesses, as there would have been no preceding human beings, and witnesses not human have, in the present day, no standing in our courts. As the case stands, however, the doctrine arose in an age when man did not trouble himself about evidence, but was content to accept his opinions on authority; and this, strangely enough, is held by many to be a strong point in its favor, it gaining, in their minds, authenticity from antiquity. It is claimed, indeed, to be sustained by divine authority, but this is a claim that has no warrant in the words of the statement itself, and one to which no form of words could give warrant. To establish it, direct and incontestable evidence from the creative power itself would be necessary, and it need scarcely be said that no such evidence exists. It is not easy, indeed, to conceive what form such evidence could take. It would certainly need to be something far more convincing than a statement in a book.

It might have been better for civilized mankind if the opening pages of Genesis had never been written, since they have played a potent part in [3] checking the development of thought. As the case now stands, the cosmological doctrines they contain can no longer claim even a shadow of divine authority, since they have been distinctly traced back to a human origin. It has been recently discovered that they are simply a restatement of the Babylonian cosmology, as given in a literary production ages older than the Bible, an epic poem of very remote date. They are, doubtless, an outgrowth of the cosmological ideas of early man, and those who accept them must do so on the basis of belief in their probability; it is no longer permissible to claim for them the warrant of divine origin.

Modern science stringently demands facts in support of any assertion, the word "faith" having no place in its lexicon. Facts are absolutely and necessarily wanting in support of the creation doctrine, and the only argument its advocates can advance is one that deals in negatives, and demands its acceptance on the ground that the opposite doctrine has not been proved. Such an argument is valueless. Disproof of one statement is never proof of another. Its effect is simply to leave both unproved, and neither, therefore, in condition for acceptance. In the present case the weight of disproof is small. The facts in support of the evolution hypothesis are multitudinous, and many of them of great cogency; the facts against it are few, and none of them absolute. It is simply [4] argued that some questions remain unsolved, and that there are facts which seem inconsistent with the Darwinian theory of development, and which no supplementary hypotheses have explained. But no advocates of evolution hold that the Darwinian theory is final. Evolution is a growing doctrine. It has been expanding ever since it was first promulgated. Various seeming difficulties have been explained away, and it is quite possible that all may disappear as investigation widens. No such arguments add any weight to the opposite view, which has not and never could have any standing in science, since it is impossible to adduce any facts to sustain it. We shall therefore dismiss it from further consideration, and proceed to state certain general facts in favor of the evolutionary hypothesis of the origin of man.

[5]

II

VESTIGES OF MAN'S ANCESTRY

When, some centuries ago, men began to find fossil remains of animals in the rocks, a severe shock was given to the prevailing doctrine of the recent creation of the earth. The adherents of the old theology made strenuous efforts to explain away this unwelcome circumstance. The shells found had been dropped by pilgrims on their way to Jerusalem; they were mineral simulations of shells; they had been created by the Deity and placed where found; they were anything but what they appeared to be, the existing evidences of a long ancient period of animal life reaching back very far beyond the assumed date of creation.

It need scarcely be said that these explanations, especially the one that God had created fossil forms to deceive man, for some incomprehensible purpose, could not long be maintained. Some of them were inconsistent with the facts, others with common sense, and in due time it was everywhere admitted that the earth is of remote duration and has been inhabited by animals and plants for untold ages. Its structure revealed its history; its annals were found to be written in the rocks; its anatomy was full of the evidences of its origin.

[6] When, not many years ago, men began to find the fossil remains of ancient structures in the body of man himself, theology was brought face to face with a problem as difficult to explain, from its special point of view, as that of the fossils in the rocks. As the latter had threatened and finally disproved the doctrine of the special creation of the earth, so the former assailed the doctrine of the special creation of man, and annihilated it in the minds of many eminent scientists. It formed a prominent argument in favor of the theory of organic evolution, and as such calls for consideration here, as a suitable groundwork for our special theme.

The structures referred to may justly be called fossil, since they present strong evidence of being the useless remains of structures which played an active part in the bodies of some former animals. A significant example of this exists in the vermiform appendix, a narrow, blind tube descending from the cæcum of man, and detrimental instead of useful, since it is the seat of the frequently fatal disease known as appendicitis. This tube, usually from three to six inches long and of the thickness of a goose quill, is occasionally absent in man, occasionally of considerable size. It is quite large, as compared with the other intestines, in the human embryo, but ceases to grow after a certain stage of development. The cæcum is extremely long in some of the lower vegetable-eating animals, and the vermiform appendix seems to be a rudiment of the [7] formerly extended portion of this organ. It is large in the anthropoid apes, especially in the orang, in which it is very long and spirally convoluted. Its survival in man as a useless and dangerous aborted organ is a powerful argument in favor of his descent from the lower animals.

In the brain of man and many of the lower vertebrates, hanging by two peduncles, or strands of nerve fibre, from the thalami, or beds of the optic nerve, is a small rounded or heart-shaped body of about the size of a pea, known as the pineal gland. It is so destitute of any evident function that Descartes, in lack of any more probable explanation of its presence, ascribed to it the noble duty of serving as the seat of the soul. Late research has been more successful in tracking this organ to its lair. It is larger in the embryo than in the adult man, still larger in some lower vertebrates, and in certain lizards has been found to exist as an eye, its parts plainly distinguishable under the microscope. It is placed in the middle of the forehead, between the other eyes, and was no doubt an active organ of vision in some ancient batrachians.

The pineal eye, as it is now named, once useful, long useless, has persisted as a fossil structure through a far extended line of development. No more convincing evidence that man gained his body through descent from the lower animals could be asked for than the survival in the human [8] brain of this wonderfully significant remnant of a formerly useful organ. Like various other vestiges of ancient organs, it is not only useless but detrimental. It occasionally enlarges and becomes the seat of large and complicated tumors, which may cause death by their compression of the brain.

Two other structures common to most of the vertebrate animals exist in man, though they render him little or no service. These are the thymus and thyroid glands, apparently vestigial structures. The thymus gland attains a considerable development in the embryo and shrinks away to the merest vestige in the adult. It begins to form early in the embryo life as an epithelial ingrowth from the throat, and extends from the neck into the chest. It continues to grow after birth, but later begins to shrink and nearly disappears in the adult.

The thyroid gland has a somewhat similar origin, it beginning as an ingrowth from the lower section of the pharynx and extending down to the lower part of the neck. It subsequently loses its connection with the pharynx, and in adult life is a bilobed structure on either side of the windpipe. Like the thymus it is a ductless gland, abundantly supplied with blood-vessels, and possesses a vast number of small cavities, lined with cells and containing an insoluble jelly. So far as appears, both these glands are useless, or nearly so, to man; or if the thyroid performs any useful service it is a [9] minor and obscure one. Such functions as it may have could probably be performed by some of the other organs, while it is positively detrimental as the seat of goitre. This unsightly disease is due to its enlargement, either by a great increase of its blood-vessels or a development of the capsules and increase of their contained jelly. Dr. S. V. Clevenger considers these organs to have had a branchial or respiratory origin, saying that there are many reasons for believing them to be rudimentary gills. Owen says that the thymus appears in vertebrates with the establishment of the lung as the main or exclusive respiratory organ. It is wanting in all fishes, also in the gill-bearing batrachians, siren and proteus. The thyroid appears in fishes, and Gegenbaur believes that it may have been a useful organ to the Tunicata in their former state of existence.

Dr. Clevenger, in the *American Naturalist* for January, 1884, points out another curious structure in man, whose significance does not seem to have been previously observed. This is a strange and striking fact relating to the formation of the veins. It is well known that these organs possess valves, which permit the free upward flow of the blood toward the heart, but resist its descent through the action of gravity, in this way aiding its return from the extremities. The rule holds good throughout the quadrupeds that the vertical veins possess valves, while they are absent from the horizontal veins, in which they would be of no utility. But the singular fact exists that in the human trunk the valves occur in the horizontal and are absent from the vertical veins. In other words, they exist where they are useless for their apparent purpose and are absent where they would be useful.

The only conclusion that can reasonably be drawn from this strange fact is that we are here dealing with a fossilized structure, a functionless survival. It leads irresistibly to the inference that man has descended from a quadruped ancestor, and that when his body took the upright position the structure of the veins, not being seriously detrimental, remained unchanged. Those which had been vertical became horizontal, and retained their now useless valves; those which had been horizontal became vertical, and remained destitute of valves. The veins of the arms and legs, vertical in both forms, retained their valves.

Dr. Clevenger points out that the intercostal veins, which carry blood almost horizontally backward to the azygos veins and which would run vertically upward in quadrupeds, possess valves. These are not only useless to man, but when he lies upon his back they are an actual hindrance to the free flow of the blood. In like manner, the inferior thyroid veins, whose blood flows into the innominate, are obstructed by valves at the point of junction.

We quote from him as follows: "There are two pairs of valves in the external jugular and one pair in the internal jugular, but in recognition of their uselessness they do not prevent regurgitation of blood nor liquids from passing upward. An apparent anomaly exists in the absence of valves from parts where they are most needed, as in the venæ cavæ, spinal, iliac, hæmorrhoidal, and portal. The azygos veins have imperfect valves. Place men upon 'all fours' and the law governing the presence and absence of valves is at once apparent, applicable, so far as I have been able to ascertain, to all quadrupedal and quadrumanous animals: *Dorsal veins are valved; cephalad, ventrad, and caudad veins have no valves.*"

Of the few exceptions to this rule, he considers the valves of the jugular veins as in process of becoming obsolete, and the rudimentary azygos valves as a recent development. Valves in the hæmorrhoidal veins would be out of place in quadrupeds, but their absence in man is a serious defect in his organization, since the resulting engorgement of blood gives rise to the distressing disease known as piles. The presence of valves would obviate this.

No one can argue that this useless and, to some extent, injurious condition is a designed result of creation. There could not, indeed, be stronger evidence that man has descended from a quadruped ancestor. Dr. Clevenger points out other serious [12] results of the upright position of the body, from which quadrupeds are free. One of these is the liability to inguinal hernia, or rupture, which leads to much suffering and frequent death in man. Prolapsis uteri is another, and a third to which he particularly alludes is difficulty in parturition.

It has been suggested above that the thyroid gland may possibly be of some minor functional importance, and that the thymus is developed in the embryo sufficiently to be functional. As regards the latter, no one is likely to maintain that an act of direct creation would include the production of an organ of some slight and obscure utility to the embryo and useless in later life. The strong probability is that this gland belongs in the same category with other embryonic survivals yet to be pointed out. As regards the seeming function of the thyroid, it may be said that the surviving relic of an ancient functional organ is quite capable of varying in structure and taking upon itself a new function, of minor value, which in its absence would be left undone or be performed by some of the other organs.

A highly interesting example of this exists in the swim-bladder of the fish, which there is good reason to believe is a survival of an ancient structure used for quite a different purpose. It was originally developed, in the opinion of the writer, [1] [13] as an air-breathing organ, in a very ancient semi-amphibious class of fishes, from which the existing bony fishes have descended. When the latter resumed the gill-breathing habit, this organ lost its original function, and its subsequent history is a curious and significant one. In some modern fishes it has quite disappeared. In others it exists as a minute and useless remnant, no larger than a pea. In many it has been converted into the swim-bladder, and in this form serves a useful purpose, but varies very greatly in shape and size. Finally, in a few instances, it retains some measure of its probably original function of air-breathing. It is a fact of much significance, that those fishes without a swim-bladder do not seem to be at any disadvantage from its absence, but are able to make their way vertically through the water quite as well as those which possess this organ. The presumption, therefore, is that it is of little utility to the fish, and that its employment for this purpose is a mere resultant of its survival and character. Such an organ could never have been evolved as an aid in swimming, since its shrinkage to a useless rudiment in some cases and its complete extinction in others show that this function is in no sense a necessary one. It is there and has lost its old use, and is, in some cases, adapted to another purpose; that is all that can be said.

Man is the one hairless mammal,—or hairless [14] except on a few parts of his body. Yet the whole body is covered with a thin growth of hair, useless for any purpose of protection, and only explainable as a survival from the mammalian covering. The occasional considerable development of the hair is an indication pointing to such an origin. This applies not only to individuals, but to tribes or races, as in the instances of the Ainos of Japan and some of the Pygmies of Africa. The disappearance of the hair in man has been traced to no well established cause. Darwin's view that it may have been a result of sexual selection seems the most probable explanation. Certainly this is the case with the beard, whose absence in women shows it to be of no utility, and whose presence in man is in accord with the many structures in male animals apparently due to this form of selection.

Darwin has pointed out and explained a very curious peculiarity of the hair in man, which is absolutely inexplicable except on the theory of descent. This is the fact that the hairs on man's arms are directed toward the elbow from above and below, thus growing in opposite directions on the upper and lower arms. The same peculiarity exists in the larger anthropoid apes and in some of the gibbons, but is not found in the lower mammals. In the apes it is believed to be due to the habit of protecting the head from rain by covering it with the hands, the hairs turning so that the rain can run downward freely in both directions toward [15] the bent elbow. This is so useless in man that it can be explained only as a survival.

There are some other survivals in man of ancient structures to which a passing allusion must suffice. In man's eye is a minute membrane, the semilunar fold, which is absolutely useless in his economy. There is every reason to believe that this is the rudiment of a membrane which is fully developed in many animals, and is especially useful to birds, the nictitating membrane, or third eyelid. Again, the muscles which move the skin in many animals, especially in horses, have left inactive remnants in many parts of the human body. These are normally active only in the forehead, where they serve to lift the eyebrows, but they occasionally become active elsewhere. Thus there are some persons who can move the skin of the scalp. Darwin cites some who could throw heavy books from the head in this manner. The same may be said of the rudimentary muscles of the ear. There are persons who can move their ears in the same way as is done by the lower animals. Again, the whole external ear may be looked upon as a rudimentary structure, since it does not appear to aid the hearing in man. As regards the pointed ear of man's probable ancestor, Darwin calls attention to what seems a trace in man of the lost tip.

Carrying this consideration farther, it may be asked, Of what use are the five toes to man? Would not a solid foot have answered the purpose [16] of walking quite as well? But as survivals their presence is fully accounted for, since they are indispensable to many of the lower animals. Question may also be made of the utility of the large number of bones in the wrist and heel of man. Equal flexibility of the joint could certainly have been obtained with a smaller number of bones. It is only when these are traced back to their probable origin in the walking organs of the fish ancestor of the batrachians that their presence becomes explainable. They are apparently survivals of a very ancient structure, originated for swimming, and adapted to walking.

As regards the wrist of man, a curious prediction that a certain bone found in some of the lower animals, the *os centrale*, would be found in man has been made and verified, it being discovered as a very small rudiment in the human embryo. The tail, so common a feature in the lower animals, but absent from the higher apes and from man, has not vanished without leaving its traces. In the human embryo it is plainly indicated; and while it vanishes in man beyond the embryo stage, it is simply hidden beneath the skin, where its vertebrae are still apparent, usually three, sometimes four or five, in number. In addition to this, the muscles which move the tail have left traces of their presence, which not infrequently develop into true muscles.

In the human embryo, indeed, we find ourselves [17] in the midst of highly significant indications of man's origin. The body of man passes in its early development through a series of stages, in each of which it resembles the mature or the embryo state of certain animals lower in the stage of existence. It begins its existence as a simple cell, analogous in form to the amoeba, one of the lowest living creatures, and later assumes the gastrula form supposed to have been that of the earliest many-celled animals. From this state it progresses by successive stages, each of which has some relation in form to a lower class.

The most significant of these is that in which the embryo is closely assimilated to the fish, by the possession of gill slits. There are four of these openings in the neck of the human foetus, and they are at times so persistent that children have been born with them still open, so that fluids taken in at the mouth could trickle out at the neck, the opening being sufficient to admit a thin probe. [2] These slits are utilized in the developing embryo, one of them being devoted to an important duty, that of conversion into the external and middle ear. Thus the opening for hearing is an adaptation of what was once an opening for breathing. Occasionally an ear-like outgrowth appears on the neck, indicative of the attempt of a second slit to develop into an ear. The purpose of the gill slits is made more apparent by the [18] presence in the embryo of gill arches of the blood-vessels, like those normal to the fish. These disappear in common with the slits.

The temporary appearance of these gill slits is the strongest evidence that could well be demanded that the human embryo passes through the various stages which the adult has assumed in its long development in past time, and that one of these stages was the fish.

And these form only one of the evidences of man's origin to be found in the embryo. Another which may be mentioned is the wool-like hair which covers the foetus, and whose presence is incomprehensible except on the theory of descent. Its most probable explanation is that it appears as a passing survival of the first permanent coat of hair of the lower mammals.

In the milk teeth of man we have another useless and often annoying survival of an ancient state of the dental organs. We cannot well imagine that in any direct creation a set of temporary teeth would have been provided as preliminary to a permanent set—an utterly useless provision. But when we find that in a lower stage of animal life the old teeth are periodically succeeded by new ones, we can understand how a trace of this condition has persisted in the mammalia.

Other evidences of man's origin in the lower animals could be drawn from the phenomena of atavism, or arrest of development in parts or organs of the body. Atavism is usually confined [19] within the line of human descent, conditions appearing in many of us which belonged to some of our human ancestors a few generations, occasionally many generations, in the past. But conditions now and then appear which are abnormal to man, but which are normal to some of the lower animals. This tendency is exhibited by all organisms. In an occasional horse the long-lost stripes of the zebra-like ancestor reappear. Now and then a blue pigeon, like the ancestral form, crops up in a pure breed of domesticated birds. Even in the details of anatomy some long-vanished character suddenly appears.

Many instances of this in man might be cited, embracing various features of the muscular and other internal organs. The abnormality of club-foot may be pointed to as a reversion to the shape of the foot in the anthropoid apes. This, however, is a retention of a condition existing in the foetus of man, the foot being drawn up and the sole turned inward and upward. It is simply a passing testimony to the ancestral condition of man.

Again, we have the fact that man possesses normally only twelve ribs, one less than is found in the gorilla and the chimpanzee. This leads to the possibility that man may have lost a rib in his development, and in significant evidence of this is the fact that occasionally a thirteenth rib appears in the human framework.

The functionless organs in men are, as above [20] said, closely analogous to the fossils in the rocks, in that both point back to a period in which they were active, vital forms occupying a definite place in the long line of animal life or animal structure. The argument that God directly created the fossils is no more absurd than the one that He directly created these useless and at times detrimental organs. It is impossible to offer a reason for such a futile exercise of creative power, unless that it was intended to make it falsely appear that man arose from the world of life below him. Will any one in this age assert that God placed useless and dangerous structures in the body of man for the incredible purpose of deceiving him in regard to his origin? And will it be further asserted that the Deity placed similar stumbling-blocks to the human reason in the embryo, in order to deceive those who should extend their researches to this low level? It would be difficult to conceive of a more preposterous idea, yet there is no other escape from what seems a self-evident fact, that man is a product of evolution from the lower animals, and bears the marks of his ancestry thick upon him.

FOOTNOTES:

[1] "On the Air Bladder of Fishes." Proceedings of the Academy of Natural Sciences of Philadelphia, 1885.

[2] Sutton, "Evolution and Disease."

[21]

III

RELICS OF ANCIENT MAN

If now, instead of seeking for evidences of man's ancestry within the human body, in survivals of ancient anatomical structures, we seek for them within the crust of the earth, we find ourselves confronted with evidences of a great antiquity of the human race, partly in implements of human manufacture, partly in ancient or fossilized bones of primitive man. These indicate not only great remoteness of origin, but also a very gradual advance from the lowest stage of inventive ability to the high level now attained.

These relics of primitive man are divided by Dana into ten varieties, (1) Buried human bones; (2) stone arrow and lance heads, hatchets, pestles, etc.; (3) flint chips, left in the manufacture of implements; (4) arrow heads and other implements made of bone and deer horn; (5) bones, teeth, and shells bored or notched by human hands; (6) cut or carved wood; (7) bone, horn, ivory, or stone graven with figures, or cut into the shapes of animals; (8) marrow bones broken longitudinally to obtain the marrow for food; (9) fragments of charcoal and other indications of the use of fire; (10) fragments of pottery.

[22] Relics of the kinds above cited have been found at intervals for many years past, but their age and significance were doubted, and only within some forty years has the great antiquity of man upon the earth been generally acknowledged by scientists. The most important early find of ancient implements was made by Boucher de Perthes in 1841 and subsequently, in the high level gravels of the valley of the Somme, in Picardy, France. In deep layers of these gravels, which were deposited at a period when the river occupied a wider and higher channel than at present, he found rude flint weapons and tools, bearing plain evidences of human workmanship, and mingled with the teeth and bones of animals, both of living and extinct species. Among the bones were those of the mammoth and the hairy rhinoceros, species evidently contemporary with man, though they have long since vanished from the earth. At a somewhat earlier date, implements of men, mingled with bones of the cave-bear, cave-lion, hyena, and other species, had been found

in the caves of France and Belgium. These were frequently buried beneath deposits of stalagmite and other materials that must have taken a long time to accumulate.

The significance of these discoveries was long in forcing itself upon the attention of scientific men. Nearly twenty years passed before Boucher de Perthes could get the noted geologists of France and England to investigate the Somme gravels. [23] When they did so they were quickly convinced of the genuine antiquity of these relics, and announced it as a fact beyond question that man had lived in the Somme valley and fashioned rude implements out of flint during what was known as the Quaternary or Drift Period of geology.

The discoveries here made set men actively at work investigating elsewhere. Excavations were made in other high level gravels, caverns were carefully and minutely examined, Kent's Cavern, England, was dug out to its rock bottom, dozens of important finds resulted, and the antiquity of man was proved to extend back from thousands to tens of thousands, if not to hundreds of thousands, of years. And the coexistence of man with the animals whose bones accompanied his relics was proved by unquestionable evidence, for drawings and carved forms of these animals were found, proving incontestably that man had gazed upon their living forms. Thus the sketch of a mammoth, showing the long hair which served to protect this animal from the cold, was found engraved upon a piece of mammoth ivory, and one of a group of reindeer on a piece of reindeer horn. There were also drawings of the cave-bear, the seal, etc., and one very interesting group showing the aurochs, a number of trees, and a man with a snake apparently biting his heel. The carvings consisted of the horn handle of a dagger, cut into the shape of a reindeer, and other forms.

[24] That these relics belong to a far distant age is proved by the strongest evidence. It must suffice here to give some of the more striking of these proofs of antiquity. The flint hatchets found at St. Acheul, France, were obtained from a gravel bed which lay below twelve feet of sand and marl. On the surface was a layer of soil, in which were graves of the Gallo-Roman period, showing that it had been there for at least fifteen hundred years. The time needed for

the slow accumulation of the whole series of deposits must have been very considerable.

A much more decisive proof of antiquity is given by the position in which this and similar gravel beds lie. They are found along the sides of rivers at a height often of a hundred or two hundred feet above the flood level of the streams. When they were deposited, the rivers must have run at this elevation, so that time has since elapsed sufficient for the streams to cut down their valleys to the present depths. The streams may have formerly been of greater volume, and had superior cutting powers, and they may have been aided by the ice of the Glacial Age, yet, however we estimate, the conclusion is inevitable that the men who dropped their implements into those gravels must have lived upon the earth ages before the beginning of historical times.

The presence there of remains of animals which ages ago perished from the earth is another cir [25] cumstance indicative of high antiquity. These embrace the mammoth, — the great hairy elephant of prehistoric times, — an extinct hair-clad rhinoceros, the large and powerful cave-bear and cave-lion, the great Irish elk, and still other animals of whose existence we know only by their bones. Others, which existed in common with men of later date, are the reindeer and the musk-ox, species of which now inhabit the coldest regions of the north, and whose presence in southern Europe at that era seems to indicate a much colder climate than that of historic times.

The evidences of human antiquity here briefly presented are accompanied by indications of a gradual development of the human intellect. If man has "fallen from his high estate," he has left no traces of this high estate on his downward path. We possess abundant indications of his upward climb, we find none of a preceding descent. If we base our opinions on known facts, the theory of development is the only one that can be sustained; the doctrine of a fall is absolutely without warrant outside the pages of Genesis.

The successive stages of man's mental development, as indicated in the work of his hands, are well and clearly marked. At the lowest level we find tools and weapons of the palæolithic or old stone age, made of roughly chipped stone, rude in form, and never ground or polished. These present some evidence of gradual improvement, but [26] we must go to a higher level to find implements of a decidedly higher order, the neatly shaped and polished stone implements of the neolithic or new stone age. With the coming of these appears a much greater diversity in tools and weapons, and evidences of a growing skill in manufacture and a considerably greater power of invention. Still higher lie the deposits of the bronze age, in which metal replaces stone in human implements. Finally appears the age of iron, that in which we still remain. We need merely refer in passing to the lake-dwellings of Switzerland, with their many interesting relics of man during the later stone, the bronze, and the early iron eras; and the kitchen-middens, or refuse-heaps, of the Danish islands and elsewhere, which extend from the old stone age far down toward the historic period.

These are but a portion of the evidences of man's antiquity and his gradual progress in the arts of manufacture. Others have been found in many parts of the earth. Many of them exist in America, proving that man resided on this continent at a very distant era. When we consider that late discoveries in Babylonia appear to carry back the age of civilization and historical relics to some ten thousand years, and that semi-civilization must have extended very considerably beyond that time, the vista of man's gradual progress seems to recede interminably and the era of primitive man to stretch backward to an enormously remote [27] period. In truth, discoveries have been made which are claimed to carry man back beyond the Quaternary and into the Tertiary Period of geology, since cut and scratched bones have been found in Pliocene deposits, which some geologists of experience believe to have been the work of human hands. Still more remote are some seemingly chipped flints and bones cut in a way that suggests human action, which have been found in deposits of the very far-distant Miocene Age. The immense remoteness of this epoch and the rudeness of the work have cast much doubt on the human origin of these remains, though their authenticity as the work of man has been accepted by several competent observers, among them the able anthropologist, Quatrefages.

If we confine ourselves, however, to the conclusions regarding ancient man which are generally accepted, we must say that he has not been clearly traced back beyond the Glacial Period, though some of the relics found in the older river gravels and in the lowest cave accumulations may well be of pre-glacial age. Many geologists believe that he reached Europe as early as the extinct mammals with which he was contemporaneous there, but how far back in time this would carry his advent it is impossible to say.

Coming now to the consideration of more immediate human relics, the bones of man himself, it must be said that well-authenticated remains of [28] palæolithic or early neolithic man are not numerous. As long as man left his bones to the unaided agencies of nature, they were little likely to be preserved. Of the anthropoid apes of Europe, probably numerous in individuals, a few remains of one or two species alone survive. Of pre-glacial man none remain, but this may merely indicate that he has shared the fate of numerous other species that died out and left no trace. It was only when the growing cold drove man from the open woods to seek shelter in caves that remnants of his body were likely to be preserved, and only when a growing sense of human dignity led to the art of sepulture that the preservation of his bones became assured.

The burial art was seemingly not practised by the hunters of the river-drift period or by men of still earlier date. The only remains of primitive man known are those found in caves and rock shelters. A number of human skulls have been discovered in these situations, and in a few instances skeletons have been exhumed. In the neolithic period interment became more common and more carefully performed, and the progress of this period is marked by many remains of man, which in later times were buried in elaborately constructed stone sepulchres, sometimes massive in materials and covered by great earth-mounds.

What is meant by the Glacial Age is probably well-known to most readers, but its close relations [29] to ancient man render it important for those who are not familiar with its meaning that a passing description of it should here be given. It will suffice to say that there are found over much of the northern portions of America and Europe accumulations of clays, sands, and gravels, sometimes laid down in stratified beds, sometimes rudely piled together. In these occur blocks of stone, large and small, and other blocks, occasionally of great size, are found in isolated localities. The solid rocks which lie beneath these heaps are often scratched or polished, as if the material had been pushed over them with great force.

All geologists now believe that these accumulations were made by ice, at some remote period when a very cold climate prevailed in the northern hemisphere, and great glaciers slowly made their way southward, grinding and rending as they went, and burying the land under their mountain-like heaps, which sometimes were a mile or more in depth. In North America the glacial ice pushed southward to the 40th degree of north latitude. In Europe it extended to the Alpine region, but failed to reach the countries bordering on the Mediterranean.

The elaborate and minute investigation of the glacial deposits has made it highly probable that there were two glacial eras, two periods in which the ice pushed down far to the south, and that these were separated by a period in which the ice [30] retreated and an age of warmer weather intervened. This is known as the interglacial period. So far as can be positively ascertained, all the authentic relics of man belong to the Glacial Age. They seem first to become numerous in the interglacial period, and continue to increase and become diversified as we descend lower in time. How long ago it was that the sea of ice began its downflow over the earth it is impossible to say. Some place it back six hundred thousand or seven hundred thousand years. Some seek to bring it down to a quite recent date. It is still so uncertain and such a matter of controversy that the utmost we are able definitely to say is that it was very long ago.

While there is no positive proof that men dwelt in Europe before the coming on of the glacial chill, we have no just reason to doubt it. That he lived there during glacial times is unquestionable, and we may be very well assured that a naked tropical animal, destitute of the hairy covering of the other animals, would not have chosen that frozen period to migrate to the north. The fact that he was there during the ice age seems satisfactory evidence that he was there before that age, during the mild climate of late Tertiary times, and that—for a reason which we shall hereafter consider—he was caught there and unable to retreat, and was forced to adapt himself to the new conditions.

During the warm preceding period he probably [31] wandered as a hunter through the European forests. But with the gradual coming on of a wintry chill, as the advance of the ice began, shelter of some kind became necessary, and he sought refuge in caves. From being a forest wanderer he became a troglodyte. Everywhere in southwestern Europe we find traces of this period of man's existence. There is hardly a cave or rock shelter in that region within which he has not left his marks. He made his way to England, which was probably then connected by land with Europe, and dwelt long in its caverns. His period of cave residence, indeed, appears to have been a very extended one. While it continued, deposits many feet in depth gradually accumulated on the floors of the caverns, slowly filling them up. And that, in some cases at least, this cave residence ended a very long time ago, we are assured, for since then a great thickness of stalagmite, which is deposited with extreme slowness, has spread over the lower cave deposits and sealed them in.

It is in these caves that we find, not only the rude stone spearheads, scrapers, hammers, etc., the bone awls, borers, and other implements of palæolithic man, but the bones of man himself. And it is significant of his primitive condition that these earliest relics indicate a man of a very low grade of development, mentally far above the ape, it is true, but mentally and physically much below modern man.

[32] The most ape-like of those human remains is the famous Neanderthal skull, found in 1856 in a limestone cavern of the Neanderthal Valley, between Düsseldorf and Elberfeld, in Rhenish Prussia. The relics discovered consist of the brain cap, two femori, two humeri, and other fragments. The fragment of the skull attracted wide attention by its bestial aspect, it presenting a low, narrow and receding forehead, and an enormous thickness of the bony ridges over the eyes, like that seen in the gorilla. This skull, which was associated with remains of the cave-bear, hyena, and rhinoceros, is, with one exception, the most ape-like human relic yet found. Yet its cranial capacity is far above that of the highest apes, and is assimilated with that of Hottentot and Polynesian skulls.

It has been maintained that this is a pathological specimen, and does not represent normal man. But this theory has been disproved by the fact that other skulls of similar cranial characters are now known, indicating that the Neanderthal cranium represents a type of man, not an abnormal individual. In the Spy Cavern, in the province of Namur, Belgium, there were found, in 1886, two nearly perfect skeletons of a man and a woman, both of them with very prominent eye ridges, low, retreating foreheads, and large orbits. This was strikingly the case with the woman. The lower jaws in both were heavy, while the woman was almost destitute of a chin—a marked ape-like characteristic. The [33] tibia was shorter than in any known race and stouter than in most. Its curious feature was the articulation with the femur, which was such that to maintain the equilibrium the head and body must have been thrown forward, as is the case in the anthropoid apes.

In the cave of Naulette, near Dinant, Belgium, has been found the lower jaw of a man of decidedly ape-like aspect. Its prognathism or protrusion is extreme, and the canine teeth were very strong, while the molars were evidently large and increased in size backward, a non-human characteristic. At La Denise, in the upper Loire, France, have been found the frontal bones of a man like the Neanderthal man in type, the forehead being depressed and retreating, and the superciliary ridges large and thick. Several other skulls of this general type are known, but the above will suffice as examples.

Remains of palæolithic man of considerably higher type are not wanting. In the rock shelter of Cro-Magnon, France, were found the bones of three men, one woman, and one child, of more advanced character. These, however, are of late date and may have been early neolithic. At Engis, near Liège, Belgium, a deeply buried skull, associated with many remains of extinct animals, has been dug up, which is by no means ape-like in character. A still superior example of palæolithic man is the skeleton found in a cavern at Mentone, east of Nice, France, which represents a man six [34] feet in height, with rather large head, high forehead, and very large facial angle (85°). The cave contained bones of extinct animals, but no trace of the reindeer.

There is no occasion to speak here of the many remains of neolithic man that have been exhumed. Sparse in the early part of the age of polished stone weapons, they gradually became numerous, and merged into the human remains of late prehistoric times. The American continent is not without its relics of ancient man, the most famous of which is the Calaveras skull, found in 1886 in the auriferous gravels of Calaveras County, California, at an extraordinary depth. The miners, in excavating a shaft, passed through several layers of lava and gravel, forming a total thickness of seventy-nine feet of lava and a considerable thickness of gravel, making nearly one hundred and thirty feet in all. At this depth a skull was found imbedded in the gravel, which, if authentic, must have been overflowed by several successive thick outpours of lava in the ancient volcanic era of that region. As its authenticity is, however, still a matter of controversy, nothing further need here be said about it.

Leaving these evidences of human antiquity, we come to the most remarkable and significant of all the known relics of man, if indeed it is man, for it seems to many a link between man and the ape, — not yet human, while no longer simian. This is [35] the fossil find made by Dr. Eugene Dubois in 1891 on the banks of the Bengawan River, Java, and named by him *Pithecanthropus erectus*, he maintaining that it represents a new genus of upright animals, or even a new family. The remains found by him consisted of the upper part of a skull, a molar tooth, and a femur, possibly not belonging to a single individual, as they were somewhat separated. These were exhumed from a stratum of volcanic tufa, claimed to be of Tertiary age, but perhaps Quaternary, and lay at a depth of some forty feet beneath the surface.

The femur very closely resembles that of a human being of average size, and its shape, articulating surface, and other characters show clearly that the animal stood habitually erect. The principal significance lies in the tooth and the cranium. The former is like that of the chimpanzee in shape, but less rugose on its grinding surface. It seems to lie between the ape and the human type of dentition. The cranium has a low, depressed arch, with a very narrow frontal region and highly developed superciliary ridges. The cranial capacity was apparently about one thousand, that of man being from thirteen hundred to fourteen hundred. It is therefore said to be "the lowest human cranium yet described, very nearly as much below the Neanderthal as that is below the normal European."

Professor O. C. Marsh, in a paper on the subject in the *American Journal of Science,* for February, [36] 1895, agrees with Dr. Dubois in his view of the distinct position of this form in the animal kingdom, and says that the discoverer "has proved the existence of a new prehistoric anthropoid form, not human, indeed, but in size, brain power, and erect posture much nearer man than any animal hitherto discovered, living or extinct."

We have here given a short review of a long story. The evidences of man's former existence upon the earth are multitudinous, but any extended consideration of them is aside from our purpose, which is merely to show that the proofs of man's descent found in his physical structure are strengthened by evidences which he has left strewn behind him in his long march down the ages. Only a single conclusion can be drawn from these vestiges of man excavated from caves and gravels, namely, that they indicate a gradual and steady progression upward from a very low condition, while they nowhere give evidence of the traditional fall of man.

This is certainly the case with the relics of human workmanship. They begin with the rudest chipped stones, and very slowly improve in form and finish and become more varied, as we move upward in our search. The ground and polished stones follow, and the variety of implements considerably increases, until at length the age of metal, with its developed industries, is reached. The only seeming evidence of superior intellect [37] to be found in this gradual progress is that of the drawings and carvings left us by one group of palæolithic men. But the actual mental development indicated by these becomes problematical when we consider that similar drawings are made to-day by the Bushmen of South Africa, a race of men occupying a very low mental stage. From this fact we may fairly conclude that the possession of a simple graphic art does not necessarily indicate any considerable intellectual advance.

If we consider the remains of man himself, the few bones which mark his early pathway through time, a similar conclusion must be drawn. Beginning with Pithecanthropus, which science is yet in doubt whether to class with the apes or with men, we pass upward to the bestial Neanderthal man and his fellows of the same low type. Of the sparse remains of palæolithic man that exist, the most are of this degraded type. The cranial capacity is usually not small. They had the full brain development of man. But this simply assimilates them with the low races of existing savages, many of whom have not developed the simple art of chipping stone to form weapons and yet have brains of normal human weight.

In truth, the influences under which the development of the brain took place were not what we now call intellectual. Developing man used his mental powers actively in his dealings with the hostile forces of surrounding nature, and nearly all the [38] forces of evolution were brought to bear upon the organ of the mind, the body remaining practically unchanged. His senses became acute, his cunning and alertness high, his use of weapons skilful, but his field of mental exercise was still the outer world, and the inner world of thought remained in its embryo state. The more recent development of the mind has been in its intellectual powers, while its physical aptitudes have somewhat declined. This has not yielded any marked increase in the dimensions of the brain, but it may have had a decided effect upon the proportion of its parts, the regions of the cerebrum devoted to intellectual activity probably increasing at the expense of the motor and sensory regions, while the convolutions may have grown considerably more complicated.

[39]

IV

FROM QUADRUPED TO BIPED

In the question which now confronts us, that of the evolution of man from the lower world of animals, it is necessary first to state in what particulars he has evolved, what are the conditions which distinguish him from the lower animals. Four marked distinctions may be named: his erect attitude, with the freeing of the fore limbs from use as agents in locomotion; his employment of natural objects, instead of his bodily organs, as tools and weapons; his development of vocal language; and his great mental superiority, with the general use of the mind in his dealings with nature.

In none of these particulars does man stand quite alone; in all of them an affinity with the lower animals exists. Steps of progress in these directions have been made by many animals, though none of them have gained any considerable advance. In man's strikingly developed social habit and organization he has no close counterpart among the vertebrates, but several among the insects. And it is of much interest to find that in the highest field of man's progress, his employment of the mind in his dealings with [40] nature, he is chiefly emulated by such lowly-organized creatures as the ants and the bees.

We do not need to look far among the lower animals for the species which come nearest to man in structure and which seem to have immediately preceded him in the line of descent. We find these forms in the monkeys or apes, and especially in their highest representatives, the anthropoid apes. These possess in a partial degree all the special characteristics of man. They are social in habit; some of them are semi-erect in posture, and their fore limbs partly freed from use in locomotion; they possess some imperfect means of vocal communication; they employ the mind to some extent in place of the body; in short, they seem arrested forms on the road from brute to man, signal-posts on the highway of evolution. In physical organization their approach to man is singularly close. In anatomy man and the higher apes are in most respects counterparts of each other. The principal anatomical distinction has been considered to be in the foot, which from the opposable character of the great toe was classed by Cuvier with the hand, the apes being named Quadrumana, or four-handed, and man Bimana, or two-handed. Fuller research has shown that this distinction does not exist, the foot of the ape being found to agree far more closely with the foot than with the hand of man. Estimated according to use, the hand is, in the whole order, the special [41] prehensile organ; the foot, however prehensile it may be, is predominantly a walking organ. And the opposability of the great toe is approached in some men, who have great mobility in this organ, and can use it for grasping.

In regard to the brain, the organ of the mind, the difference between the higher apes and man is almost solely one of comparative size, the lower intelligence of the apes being indicated by the smaller size of their brains. The largest ape brain is scarcely half the size of the smallest human brain. But anatomically they are nearly identical. All the structural features of the brain are common to both, and the details are largely filled out in the anthropoid apes, the convolutions being all present and the pattern of arrangement the same. The brain of the orang may be said to be like that of man in all respects except size and the greater symmetry of its convolutions, which are less complicated with minor convolutions than in man. In truth, the difference between the brains of man and the orang is almost insignificant as compared with the difference between those of the orang and the lowest apes. Mr. E. W. Taylor, who has recently made an exhaustive study of the minute anatomy of the brain of the chimpanzee, remarks, "The similarity between the brain of the anthropoid apes and of man is one of the most singular and interesting facts of which we have knowledge."

[42] In any attempt, then, to consider the origin of man from the point of view of evolution, we are irresistibly drawn to the ape tribe as the next lower link in the long chain of development, and are led to consider the characteristics of the apes as the intermediate stage between the quadruped and the biped, the bridge crossing this great gulf in organic development. This is by no means to suggest that some one of the existing anthropoid apes is the direct ancestor of man. Such an idea has never been entertained by scientists. These animals cannot even fairly be considered as brothers to man's ancestor, but must be looked upon as more or less distant cousins, with a physical organization less favorable to high development than that

of man. Man's ancestry lies much farther back in time, and his progenitor must have been constituted differently from any of the existing large apes.

In the ape tribe we are able to trace nearly every step by which the gulf between quadruped and biped has been crossed, from the quadrupedal baboon to the nearly erect gibbon. And in seeking to follow this development through its successive stages, the first point to be considered is how the apes gained their special power of grasping, that characteristic to which they undoubtedly owe the partial freedom of their hands and their tendency to assume the erect attitude.

The most distinguishing characteristic of the [43] apes and of the nearly related lemurs has not hitherto been definitely pointed out. This is that they form the only group of strictly arboreal animals. The tree is not alone their native habitat, but they are specially adapted to it in their organs of motion, a fact which cannot be affirmed of any other animal group. If we consider, for instance, the squirrels, one of the best-known groups of tree-living animals, we find them to be members of the great order of rodents, whose native habitat is the land surface. Though the squirrels have taken to the trees, there has been no adaptive change in the structure of their limbs and feet. The same may be said of almost all tree-dwellers except the lemurs and apes. The sloth, indeed, is specially adapted in organization to an arboreal residence, but this change is individual, not tribal, this animal being an aberrant form of the ground-dwelling edentata. In the apes and lemurs, on the contrary, the ground-dwellers are the aberrant forms, stray wanderers from the host. Nearly all the species live in trees, to which they are specially adapted by the formation of their feet. It remains to inquire how this deviation in structure arose, what were the steps of development of the grasping foot and hand, the special characteristic of this group.

In considering this question, the first fact to appear is that the apes and lemurs are plantigrade animals. Their natural tendency is to walk on the sole of the foot, a habit which few other tribes of [44] animals possess. Most of the larger animals walk on the knuckles or the toes, and develop claws or hoofs, but the ancestral form of the ape, ages in the past, was doubtless a sole-walking quadruped, its toes apparently provided with nails instead of claws. What the story of this very ancient quadruped was we are quite unable to say. It may, in the exigencies of existence, have come to a parting of the ways; a section of the group, drawn by a love of fruit, developing the climbing habit; the remaining section continuing on the ground and following a separate line of evolution. Perhaps only a single species took to the trees; for it is quite possible for a single form, in a new and advantageous habitat, to vary in time into a great number of species.

Of all this we can know nothing: but of one thing we may feel assured, which is that the plantigrade foot is the only one that could have developed into a grasping organ; such a development being impossible to the digitigrade or the hoofed animals. One can readily see how the habit of walking on the sole might tend to a spreading of the toes, in order to obtain a wider and firmer footing. And it is equally easy to see how a free and wide motion in the great toe would aid in this result. The animal may have been at first light in weight and able to support itself on its unchanged foot, but as it increased in size and weight it would need a firmer grasp, and the final result of spread [45] ing its toes for this purpose may well have been the opposable great toe.

It must be borne in mind, in this consideration, that the apes differ from the other tree-dwellers in being destitute of claws. The squirrels, the opossums, and other arboreal animals have sharp claws, by whose aid they can easily cling to the surface of the bark-covered boughs. The nails of the apes are incapable of affording them this service, and it is not easy to perceive how a foot like theirs could become adapted to locomotion in the trees otherwise than by the gaining of mobile action and grasping power in the toes.

The existing habits of the ape tribe lead us to the conclusion that the ancestral animal may have soon begun to seek support from upper limbs. The plantigrade foot is one capable of readily curving into an organ of support, and in the case of the forefoot the toes would tend to spread and gain flexibility of motion, and the first toe to become opposable to the others and yield a more complete grasping power. It does not seem difficult to comprehend, from this point of view, how the feet of a five-toed plantigrade animal may in time have developed into grasping organs, since there would be required only an increased flexibility of the joints, and a wider and fuller movement of the great toes. That such a change took place in this instance the facts appear to indicate, the most simple and probable explanation of the development of the grasp [46] ing power in the hands and feet of the ape being seemingly that given above.

The relation of the lemurs to the apes is not clearly defined. It may be an ancestral one, or the two animals may represent distinct lines of descent. In the latter case we would have two lines of animal evolution in which the grasping power was gained and adaptation to arboreal life completed. Whatever their relationship, they both possess the opposable thumb as the hall-mark of their arboreal habitat, and whenever found walking on the ground they may be looked upon as estrays from their native place of residence.

Once the grasping power was gained, the first step of change from the quadrupedal to the semi-erect attitude was completed. The process may have begun in the effort to fit the sole of the foot to the rounded surface of boughs; or its first stage may have been in the seizing of overhead branches with the flexible hand; or both influences may have acted simultaneously. We see the result only, we cannot trace the exact process; but we have as an outcome the adoption of a method of locomotion different from that of all other tree-dwellers, the forefoot developing into the hand with its opposable thumb, and the hindfoot gaining a similar grasping power in the toes.

The power of walking on a lower limb and grasping an upper one once attained, a succeeding step in evolution quickly appeared, and one [47] of prime importance to our inquiry. The animal had ceased to be in a full sense a quadruped, while not yet a biped, and a variation in the length of its limbs was almost sure to take place. This is an ordinary result when animals cease to walk on all fours. In the leaping kangaroo and jerboa a shortening of the arms and lengthening of the legs appear. Here the arms are relieved from duty and a double duty is laid on the legs, with the consequence stated. In the ancient dinosaurian reptiles, upright walkers, the same was the case. Those varied from quite small to very large animals, but in all known instances the fore limbs were greatly reduced in size. A similar condition may be seen in the birds, the bones of whose forelimbs have largely aborted from lack of employment as walking organs.

In the case of the apes and lemurs, while a similar effect has taken place, an interesting difference appears, due to the difference in conditions. In these animals the fore limbs are not freed from duty as organs of locomotion. In many cases, on the contrary, they have an extra duty put upon them, with the result that they have grown longer instead of shorter. Very likely these animals differed considerably in the past, as they do to-day, in the degree of use of their legs and arms. Many of them walk in the quadruped manner, either on the ground or in trees. Others make much use of their hands and arms in grasping and swinging [48]. Great differences in the use of the arms and legs may have arisen in different species. In some, the legs may have been mainly trusted to for support, and the hands used for steadying. In others the arms may have been the chief locomotive organs and the feet have given steadiness. Here the legs may have grown the longer, there the arms, the limbs developing in accordance with their degree of employment. In the lower monkeys and the lemurs, the bones of the pelvis are altogether quadrupedal in character. This is not the case in the higher forms, and in the highest apes the pelvic bones approach those of man.

Highly interesting examples of these varied results may be seen in the existing anthropoid apes. In all of them it would appear that the arm was a prominent factor in locomotion, for in each instance it is longer than the leg, — but it differs in proportional length in every instance. It is shortest in the chimpanzee, somewhat longer in the gorilla, still longer in the orang, and remarkably long in the gibbon. In all these instances the fact that the arms exceed the legs in length indicates that they must have played a large and important part in the work of locomotion, and especially so in the case of the gibbon. It is well known, in fact, that the gibbons progress very largely by the aid of their arms, swinging from limb to limb and from tree to tree with extraordinary strength and facility. The legs lend their [49] aid in this, but the arms are the principal organs of motion, and seem to have developed in length accordingly.

As regards the other anthropoid species, Wallace's observations on the habits of the orang are of interest. This animal usually walks on all fours on the branches in a semi-erect crouching attitude, but our naturalist saw one moving by the use of its arms alone. In passing from tree to tree the arms come actively into play. The animal seizes a handful of the overlapping boughs of the two trees and swings easily across the intervening space. While seeming to move very deliberately, its actual speed was found to be about six miles an hour.

The organization of man, as he now exists, shows an interesting and important deviation from that of the manlike apes, and one which serves as strong evidence that none of these apes occupied a place in his line of descent. This is that he is a long-legged and short-armed animal, a condition the reverse of that seen in the anthropoid apes. While man's hands reach barely to the middle of the thigh, those of the chimpanzee reach below the knee, of the gorilla to the middle of the leg, of the orang to the ankle, and of the gibbon to the ground. All these apes have short legs and long arms. Man, on the contrary, has long legs and short arms.

The natural presumption from this interesting [50] fact is that man's ancestor, which we may provisionally call the man-ape, differed essentially in its mode of progression from the other apes. The smaller forms of these usually move on all fours in the trees, though the arms are always ready for a swing or a climb. The anthropoid apes also show a tendency to a similar mode of progression, though with a difference in their mode of walking, which, as we shall see later on, is never that of the quadruped. As for the man-ape, it may have originally walked in the same manner as the related species, if we surmise that the variation in the length of the limbs was a subsequent development. Certainly after its limbs attained the proportions of those of man, its facility of swinging from tree to tree must have been diminished, while it would have found it inconvenient to move in the crouching attitude of the orang and its fellows. Its easiest attitude must then have been the erect one, and its motion a true biped walk, not the swinging and jumping movement of the other anthropoids. In short, the development of man's ancestor into a short-armed animal, however and whenever it took place, could not but have interfered seriously with its ease of motion in the trees.

Though this change may have begun in the trees, it probably had its full development only after the animal made the ground its habitual place of residence.

It is of interest to find that all the existing large [51] apes are arboreal, the gorilla being the least so, probably on account of its weight. Though they all descend at times to the ground, their awkward motion on the surface shows them to be out of their element, while they move with ease and rapidity in the trees. The organization of man renders it questionable if his primeval ancestor was arboreal to any similar extent. The indications would seem to be that it made the ground its habitual place of residence at an early period in its history, and that the result of this new habit and of its erect attitude was a change in the relative length of its limbs.

That this animal dwelt mainly in trees in the first stage of its existence, and possessed a powerful grasping power in its hands, we have corroborative evidence in recent studies of child life. The human infant, in its earliest days of life, displays a remarkable grasping power, being able to sustain its weight with its hands for a number of seconds, or a minute or more, at an age when its other muscles are flabby and powerless. It appears in this to repeat a habit normal to the ancestral infant, an instinct developed to prevent a fall from its home among the boughs.

Yet it is doubtful if the man-ape long remained a specially arboreal animal. The varied length of arm in the anthropoid apes was doubtless of early origin, and in all probability man's ancestor had originally a shorter arm than its related species. [52] If so, this must have rendered it less agile in trees than other forms. If we could see this ancient creature in its arboreal home, we should probably find it more inclined to stand erect than the other apes, walking on a lower limb, and steadying itself by grasping an upper limb. This would be a more natural and easy mode of progression to a short-armed animal than the crouching attitude of the orang or the swinging motion of the gibbon, and its effect would be to make the erect attitude to a large extent habitual with this animal.

In short, man's ancestor may have become in considerable measure a biped while still largely a dweller in the trees, and to that degree set its arms free for other duties than that of locomotion. Like the other apes, it probably often descended to the ground, where its habit of walking erect on the boughs rendered the biped walk an easy one, or where this habit may have been originally acquired. While this is conjectural, it is supported by facts of organization and existing habit, and for the reasons given it seems highly probable that the ancestor of man took to a land residence at an early period in its history, climbing again for food or safety, but dwelling more and more habitually on the earth's surface. Even at this remote era it may have become essentially human in organization, its subsequent changes being mainly in brain development, and only to a minor extent in physical form and structure.

[53] Fossil apes have not been found farther back than the Miocene Age of geology. It is quite probable, however, that they may yet be found in Eocene strata, since examples of their highest representatives, the anthropoid or manlike apes, have been found in Miocene rocks. The fact that these large apes are now few in number of species, is no proof that many forms of them may not have formerly existed, and among these we may class the ancestor of man.

[54]

V

THE FREEDOM OF THE ARMS

Man's ancestor is by no means the only form of ape that has made the earth's surface its place of residence. The baboon is one example of a number of forms that dwell habitually upon the ground, though they have not lost their agility in climbing. But these species have returned to the quadruped habit, to which the equal length of their limbs adapts them. All the anthropoid apes dwell to some extent upon the ground, but these can neither be called quadrupeds nor bipeds, their usual mode of progression being an awkward compromise between the two. The same may be said of one of the lemurs, the propithecus, the only member of its tribe that attempts to move in the erect attitude. It does not walk, however, but progresses by a series of jumps, its arms being held erect, as if for balancing.

Of the apes, though many can stand upright, the gibbon is the only one that attempts to walk in this position. This is a true walk, though not a very graceful one. The animal maintains a fairly upright posture, but walks with a waddling motion, its body rocking from side to side. Its soles are [55] placed flat on the ground, with the great toes spread outward. Its arms either hang loosely by its side, are crossed over its head, or are held aloft, swaying like balancing poles and ready to seize any overhead support. Its walk is quickly changed to a different motion if any occasion for haste arises. At once its long arms are dropped to the ground, the knuckles closed, and it progresses by a swinging or leaping motion, the body remaining nearly erect, but being swung between the arms.

None of the other anthropoid apes ever walk erect, though they assume at times the upright posture. But though they use all their limbs as walking organs, they show no tendency to revert to the habit of the quadrupeds. Their motion is like that of the gibbon when in haste, a series of jumps or swings between the supporting arms. The shortness of their arms, however, prevents them from standing erect, like the gibbon, in doing this; and they bend forward to a degree depending on the length of their arms, the chimpanzee the most, the orang the least.

As a rule, the flat sole of the foot is set on the ground, with the toes extended, as in man, but the toes are sometimes doubled under in walking. The orang rarely touches the ground with the sole or the closed toes, but walks on the outer edge of the foot, the feet being bent inward as if clasping the rounded sides of a bough. The other species have a tendency in the same direction, the legs [56] being bowed and the gait rolling. In using the hands in walking, the closed knuckles are usually placed on the ground, though occasionally the open palm is employed. The whole movement of these animals is strikingly awkward, and goes to indicate that there can be no satisfactory compromise between life in the tree and on the ground.

The significant fact in these attempts to walk is that none of the anthropoid apes show any inclination to revert to the quadruped habit. Their attitude is in all cases an approach toward the erect one, which posture is attained by the gibbon. The arms are used not as walking but as swinging organs. Evidently their mode of life in the trees has overcome all tendency toward the quadruped motion in these apes and developed a tendency toward the biped. But none of them have gained the muscular development of the leg known as the calf, nor an adjustment of the joints to the erect attitude, since none but the gibbon walks erect, and it does so only at occasional intervals.

The conclusion to be derived from all this is that the man-ape was in its early days much more truly a biped than are any of the species named. Like them, it had no tendency to revert to the quadruped habit. The shortness of its arms was unsuited to this, while rendering it impossible for the animal to progress in the semi-erect, swinging fashion of the other anthropoid apes. As a result of its bodily formation, it may have begun to walk erect at a [57] very remote date, with a consequent straightening of the joints and muscular development of the legs. When this condition was fully attained, it was practically a man in physical conformation, though mentally still an ape, and with a long development of the brain to pass through before it could reach the human level of mind.

The far-reaching conclusions here reached are all based on one important fact, the shortness of man's arms as compared with the disproportionate length of arm in the anthropoid apes. This, for the reasons given, rendered the adaptation of the man-ape to life in the trees inferior to that of the long-armed apes; while, as has just been said, it unfitted it to walk on the ground either as a quadruped or in the jumping method of its fellow anthropoids. In short, the biped attitude was much the best suited to its organization and the one it was most likely to assume. This once adopted as its habitual posture, efficiency in walking would be gained by practice.

When once this animal became a ground walker, its facility of motion in the trees was in a measure lost. When the feet became accustomed to the flat surface of the ground, they became less capable of grasping the rounded surface of the bough. Fitness to the one situation entailed loss of fitness to the other. The feet of the apes can clasp the bough firmly, by curving around its opposite sloping sides, and to this these animals doubtless owe [58] their bowed legs and their disposition to walk on the outer edge of the foot. This disposition the man-ape lost as its foot fitted itself to the surface of the ground. It was probably retained in a measure by the young, after it had been lost by the mature form, and is still manifested in the position of the foot in the human embryo.

These considerations bring us to an important question: Why did the man-ape gain a length of arm not the best suited to its arboreal habitat? Why, in fact, do changes in physical structure ever take place? How does an animal succeed in passing from one mode of life to another, when during the transition period it is imperfectly adapted to either, and therefore at a seeming disadvantage in the struggle for existence? The study of animal development has given rise to certain difficult problems of this character, some of which have been solved by showing that the supposed disadvantage did not arise, or that it was balanced by some equal advantage. In this way a considerable gap in life conditions has perhaps occasionally been crossed. Small gaps have doubtless been frequently passed over in the same manner.

In the case of the anthropoid apes, we perceive a considerable variation in the length of the arms, from the very long arms of the gibbon to the comparatively short ones of the chimpanzee. These differences are probably the result of some difference in their life habits, and accord with the [59] possibility of a still shorter arm in the man-ape. There is, however, some reason to believe, as we shall show later on, that the arm of this animal was longer and the leg shorter than in man himself, their comparative length perhaps not differing greatly from that of the chimpanzee. Aside from all other considerations, the use of the legs as the sole organs of locomotion could not well fail to produce this result, the legs growing longer and stronger in consequence of the increased duty laid upon them, and the arms growing shorter and weaker through their release from duty in locomotion. The case does not differ in character from those of the dinosauria and the kangaroos, in both of which instances a release of the arms from duty in walking was followed by a considerable decrease in length and strength, while the legs grew proportionally stronger.

If any disadvantage attended the shortening of the arms of the man-ape, to the extent that this may have taken place in the tree, it was probably correlated with some advantage. In the various instances of short-armed animals cited this appears to have been the case, and it was probably so in man's ancestral form. While the hands continued useful in grasping and enabling the animal to maintain its place on the boughs, they may have been gradually diverted to some other service, with the result that the animal found the tree less desirable than before as a place of residence and [60] sought the ground instead. This would be particularly the case if the new duty was one best exercised upon the ground.

Shall we offer a suggestion as to this new use? Such changes are usually the result of some change of habit in the animal, frequently one that has to do with its food. Change of diet or of the mode of obtaining food is the most potent influencing cause of change of habit in animals, and the one that first calls for consideration.

The apes are frugivorous animals, though not exclusively so. Carnivorous tendencies are displayed by many of them. They rob birds' nests of their eggs and young, they capture and devour snakes and other small animals. In zoölogical gardens monkeys are often observed to catch and eat mice. It is evident that many of them might readily become carnivorous to a large extent under suitable conditions. The large apes are usually frugivorous, but some of them eat animal food. This is the case with both the chimpanzee and the gorilla. The latter, while living usually on fruit and often making havoc in the sugar-cane plantations and rice-fields of the natives, also eats birds and their eggs, small mammals and reptiles, and is said to devour large animals when found dead, though it does not attempt to kill them for food. The young gorilla which was kept in captivity at Berlin became quite omnivorous in its diet.

With all this readiness to eat animal food, none [61] of the existing apes are carnivorous to any large extent, but the fact of this inclination makes it not improbable that some of the apes of the past may have been much more so. It is quite within the limits of probability, for instance, that the man-ape at an early date became omnivorous in its diet. Its change in structure may well have been the result of a decided change in diet, such as that from fruit to flesh food. Such a radical change as that from vegetable to animal food would certainly demand a more active employment of the arms as agents in capture. Fruits and nuts wait to be pulled; animals must be caught before they can be eaten. The former is an easy matter to an arboreal animal; the latter might prove a difficult one, especially if large animals were to be captured.

In short, the pursuit and capture of any of the larger animals for prey could not fail to modify to a great degree the use of the arms. Their employment in locomotion would interfere seriously with their utility in this direction. To succeed in capturing nimble prey by an animal with the ape form of hands a considerable freedom of the arms would be necessary, and the feet would have to be mainly, if not wholly, depended upon for motion. The ape has not the sharp claws of the carnivora with which to seize and hold its prey. It must have been obliged to use its palms for this purpose, and this it could not well have done unless they were free in their action.

[62] It is conceivable, indeed, that the man-ape may have run down its prey, or sprung upon it from covert, and seized it with the hands, but there is good reason to believe that this was not its mode of capture. The organization of the ape tribe gives it a characteristic action which is not to be found in any other group of the vast animal kingdom, that of handling and throwing missiles. In this it necessarily stands alone, since no other animal has a grasping palm.

The power is one of prime importance, for without it we cannot perceive how man could ever have emerged from the general animal kingdom. The use of missiles is by no means uncommon with the monkeys. We cannot safely accept the story that American monkeys will throw cocoanuts from tree-tops at those who hurl stones at them from below, from the fact that the cocoanut seems too heavy and too firmly fixed to its support for the strength of those small species, but it is not uncommon for them to throw lighter objects. Yet in doing this they usually seem to have no idea of aim, but toss the missile aimlessly into the air. Of the large apes, the orang will break off branches and fling them at its tormentors, or will throw the thick husks of the durian fruit, but with similar lack of aim. The most skilful in this exercise are some species of baboons, which can hurl branches, stones, or hard clods with much dexterity.

It is of interest to find existing apes availing [63] themselves of their grasping power in this manner, since it leads us irresistibly to the conclusion that the man-ape may have done the same thing. The species which use missiles fail to take aim for two reasons, one that they employ them only occasionally, often in imitation of human action, the other that their arms are ill suited to this motion from their constant employment in another duty. In the case of the man-ape we may justly look for a more effective result, since if the arms became relieved from duty in locomotion they were free to gain facility of action in other directions.

If in addition to this the man-ape began to use missiles with a definite purpose in view, that of striking down animal prey, so that the use of such weapons became habitual instead of occasional, it would soon gain some power of aim and a growing strength and skill in the throwing motion. It is quite probable, also, that an early use of weapons was in the form of clubs, which were retained in the grasp to strike down the prey when overtaken. In this case, we may imagine our primitive biped running swiftly after its prey, club in hand, striking at it when within reach; or, if it should prove too swift, hurling the club or a stone through the air with the hope of bringing it down in this manner. Such a flinging action, if now and then successful, would be likely soon to become habitual; while the arm would grow accustomed to this new motion, and attain skill in taking aim. We may [64] reasonably infer, also, that the club would be used for defence as well as for offence, in case the man-ape were in its turn pursued by larger animals. Instead of fleeing to the nearest tree, it might now stand its ground and beat off its enemy.

All must admit the probability, in a large tribe of animals with grasping power in their hands, and in the habit of using missiles occasionally, of one or more species coming to use them habitually. All the anthropoid apes are certainly intelligent enough to do this, if it should prove advantageous to them. Its principal advantage, however, would seem to be to a species that became largely carnivorous and needed to capture running or flying prey.

The habit of using implements is one of supreme importance in animal evolution. To it we owe man as he exists to-day. While animals confined themselves to their natural weapons of teeth and claws, their development must have remained a very slow one and been confined within narrow limits. When they once began to add to their natural powers those of surrounding nature, by the use of artificial weapons, the first step in a new and illimitable range of evolution was taken. From that day to this, man has been occupied in unfolding this method, and has advanced enormously beyond his primal state. A crude and simple use of weapons gave him, in time, supremacy over all the lower animals. An advanced use of weapons [65] and tools has given him, in a measure, supremacy over nature herself, and raised him to a stage almost infinitely beyond that of the animal which trusts solely to teeth and claws.

So far as we know, only one of the innumerable species of animals attained this development; unless, indeed, the various races of men had more than one ape ancestor. For the appearance of man there became necessary, first, the development of an order of animals with power of grasp in their hands; and, second, the development of one or more biped species, with hands freed from duty as walking organs and capable of use in other directions. A third necessity was very probably the exchange of the frugivorous for the carnivorous habit, which would act as a predisposing agency in inducing the animal to desert the tree for the ground, and to employ weapons in the chase. The final result of all this would be an erect, walking, and running animal, with arms and hands quite free from their old duty, except during an occasional return to the tree, and with the necessary straightening of joints and development of supporting muscles.

What has been advanced above is, no doubt, largely a series of assumptions and conjectures, few of which are sustained by known facts. But as the matter stands, no other method of dealing with it can be adopted, since the facts in the case have in great part vanished. What we know positively [66] is that man exists, and that in physical structure he is very closely related to the anthropoid apes. What we have excellent reason to feel assured of is that man has descended from the lower animals, and in all probability from an ape-like ancestor. We know that one or more species of anthropoid apes have become extinct, and can reasonably conjecture that one ancient species became modified into the form of man. We know that human remains have been found that, to some small extent, fill the gap between man and the ape. Correlative evidence exists in the variations in length of limb in the existing anthropoids, their efforts to walk upright, their varied degree of dependence upon the arms for locomotion, and the occasional use of missiles by these and lower forms. To these may be added the carnivorous tastes shown by many members of the ape family, with the indication that more decided carnivorous habits might readily be assumed.

Taking the stand that such a partly carnivorous anthropoid ape, biped in structure, appeared and made the ground its usual place of residence, we find ourselves on the direct trail of man. Long ago as this may have been, and far and difficult as was the journey to be made, the way was thenceforth straight and well-defined. Such an animal, living largely on animal food, and using weapons superior to its natural ones in the capture of prey, was essentially a man, however low may [67] still have been its level of intelligence. Its feet were firmly fixed upon the upward track, and only time and stress of circumstance were needed to carry it upward to the high level of civilized man.

We may, indeed, go further than this. We are in a measure justified in saying what this man-ape was like, this creature which had left its early home in the trees and began to walk upright upon the earth, pursuing the larger animals and capturing them for food. It was probably much smaller than existing man, little if any more than four feet in height and not more than half the weight of man. Its body was covered, though not profusely, with hair, the hair of the head being woolly or frizzly in texture, and the face provided with a beard. The complexion was not jet black, like the typical negro, but of a dull brown hue, the hair being somewhat similar in color. The arms were lank and rather long, the back much curved, the chest flat and narrow, the abdomen protruding, the legs rather short and bowed, the walk a waddling motion, somewhat like that of the gibbon. It had small, deep-set eyes, greatly protruding mouth with gaping lips, huge ears, and in general a very ape-like aspect. Our warrant for this description of man's ancestor must be left for a later portion of our work. We shall only say here that it is based on known fact, not on fancy.

[68]

VI

THE DEVELOPMENT OF INTELLIGENCE

The full adoption of the erect attitude gave the ancestor of man an immense motor supremacy over the lower animals, for it completely released his fore limbs from duty as organs of support and set them free for new and superior purposes. In all the animal kingdom below man there exists but a single form that emulates him in this possession of a grasping organ which takes no part in walking or in other modes of locomotion. This is the elephant, whose nose and upper lip have developed into an enormous and highly flexible trunk, with delicate grasping powers. The possession of this organ may have had much to do with the intellectual acumen of the elephant. Yet it is far inferior in its powers to the arm and hand of man; while the form, size, and food of the elephant stand in the way of the progress which might have been made by an animal possessed of such an organ in connection with a better suited bodily structure.

For a period of many millions of years the world of vertebrate life continued quadrupedal, or where a variation from this structure took place [69] the fore limbs remained to a large extent organs of locomotion. Finally a true biped appeared. For a period of equal duration the mental progress of animals was exceedingly slow. Then, with almost startling suddenness, a highly intellectual animal appeared. Thus the coming of man indicated, in two directions, an extraordinary deviation from the ordinary course of animal development. Both physically and mentally evolution seemed to take an enormous leap, instead of proceeding by its usual minute steps, and in the advent of man we have a phenomenon remarkable alike in the development of the body and the mind.

So far our attention has been directed to the evolution of the human body, now we must consider that of the human mind. In seeking through the animal kingdom for the probable ancestor of man in his bodily aspect, we were drawn irresistibly to the ape tribe, as the only one that made any near approach to him in structure. In considering the case from the point of view of mental development we find a similar irresistible drawing toward the apes, as the most spontaneously intelligent of the mammalia. While many of the lower animals are capable of being taught, the ape stands nearly alone in the power of thinking for itself, the characteristic of self-education.

Innumerable testimonials could be quoted from observers in evidence of the superior mental powers of the apes. Hartmann says of them that [70] "their intelligence sets them high above other mammals," and Romanes that they "certainly surpass all other animals in the scope of their rational faculty." It is scarcely necessary here to give extended examples of ape intelligence. Hundreds of instances are on record, many of them showing remarkable powers of reasoning for one of the lower animals. The ape, it is true, is not alone in its teachableness. Nearly all the domestic animals can be taught, the dog and the elephant to a considerable degree. And evidences of reasoning out some subject for themselves now and then appear in the domesticated species; but these are rare instances, not frequent acts as in the case of the apes.

The apes, indeed, rarely need teaching. They observe and imitate to an extent far beyond that displayed by any others of the lower animals, and the more remarkable from the fact that in nearly every instance the animals concerned began life in the wild state, and had none of the advantages of hereditary influence possessed by the domesticated dog and horse. Among the most interesting examples of spontaneous acts of intelligence of the ape tribe are those related by Romanes, in his "Animal Intelligence," of the doings of a cebus monkey, which he kept for several months under close observation in his own house. Instead of selecting general examples of ape actions, we may cite some of the doings of this intelligent creature.

[71] The cebus did not wait to be shown how to do things, but was an adept in devising ways to do them himself. He had the monkey love of mischief well developed, and not much that was breakable came whole from his hands. When he could not break an egg cup by dashing it to the ground, he hammered it on the post of a brass bedstead until it was in fragments. In breaking a stick, he would pass it down between a heavy object and the wall, and break it by hanging on its end. In destroying an article of dress, he would begin by carefully pulling out the threads, and afterward tear it to pieces with his teeth. His nuts he broke with a hammer precisely as

a man would have done and without being shown its use. Ridicule was not pleasant to him; he strongly resented being laughed at, and would throw anything within reach at his tormentor and with a skill and force not usual with monkeys. Taking the missile in both hands and standing erect, he would extend his long arms behind his back and hurl the article by bringing them forcibly forward.

If any object he wanted was too far away to reach, he would draw it toward him with a stick. Failing in this, he was observed to throw a shawl back over his head, and then fling it forward with all his strength, holding it by two corners. When it fell over the object, he brought this within reach by drawing in the shawl. In his gyrations, the [72] chain by which he was fastened often became twisted around some object. He would now examine it intently, pulling it in opposite ways with his fingers until he had discovered how the turns ran. This done, he would carefully reverse his motions until the chain was quite disentangled.

The most striking act of intelligence told of this creature was his dealings with a hearth-brush which fell into his hands, and of which the handle screwed into the brush. It took him no long time to find out how to unscrew the handle. When this was achieved, he at once began to try and screw it in again. In doing so he showed great ingenuity. At first he put the wrong end of the handle into the hole, and turned it round and round in the right direction for screwing. Finding this would not work, he took it out and tried the other end, always turning in the right direction. It was a difficult feat to perform, as he had to turn the screw with both hands, while the flexible bristles of the brush prevented it from remaining steady. To aid his operations he now held the brush with one foot, while turning with both hands. It was still difficult to make the first turn of the screw, but he worked on with untiring perseverance until he got the thread to catch, and then screwed it in to the end. The remarkable thing was that he never tried to turn the handle in the wrong direction, but always screwed it from left to right, as if he knew that he must reverse the original motion. [73] The feat accomplished, he repeated it, and continued to do so until he could perform it easily. Then he threw the brush aside, apparently taking no more interest in that over which he had worked so persistently. No man could have devoted himself more earnestly to learn some new art, and become more indifferent to it when once learned. These are a few only of the many acts of intelligence observed by Mr. Romanes in the doings of this animal. They will suffice as examples of what we mean by spontaneous intelligence. The cebus did not need to be shown how to do things; it worked them out for itself much as a man would have done, performing acts of an intricacy far beyond any ever observed in other classes of animals in captivity. It may be said further that the displays of spontaneous intelligence shown by dogs, cats, and similar animals have usually been intended in some way for the advantage of the animal; few or none are on record which indicate a mere desire to know without ulterior advantage; no persevering effort, like that with the brush, which is purely an instance of self-instruction.

Examples of intelligence of this advanced character could be cited from observation of monkeys of various species. The anthropoid apes have not been brought to any large extent under observation, but are notable for their intelligence in captivity. It is not easy to observe them in a state of nature, and nearly all we know is that the orang [74] makes itself a nightly bed of branches broken off and carefully laid together, and is said to cover itself in bed with large leaves, if the weather is wet. The chimpanzee has a similar habit, and the gorilla is said to build itself a nest in which the female and the young sleep, the old male resting at the foot of the tree, on guard against their dangerous foe, the leopard.

It is the young animals of these species which are the most social and docile and most approach man in appearance. As they grow older, their specific characters become more marked. Fierce and sullen as is the old gorilla, the young of this species is playful and affectionate in captivity and is given to mischievous tricks. The one that was kept for a time in Berlin showed much good-nature, playfulness, and intelligence, and some degree of monkey mischievousness. It was very cunning in carrying out its plans, particularly in stealing sugar, of which it was very fond.

The chief examples of anthropoid intelligence are told of the chimpanzee, which has been most frequently kept in captivity. It is usually lively and good-tempered and is very teachable. Some of the stories of its intelligence may be apocryphal, as those told by

Captain Grandpré of a chimpanzee which performed all the duties of a sailor on board ship, and of one that would heat the oven for a baker and inform him when it was of the right temperature. But there are authenticated stories [75] of chimpanzee intelligence which give it a high standing in this respect among the lower animals.

The emotional nature of the ape is also highly developed. It displays an affection equal to that of the dog, and a sympathy surpassing that of any other animal below man. The feeling displayed by monkeys for others of their kind in pain is of the most affecting nature, and Brehm relates that in the monkeys of certain species kept under confinement by him in Africa, the grief of the females for the loss of their young was so intense as to cause their death. More than once an ardent hunter has seen such examples of tender solicitude among monkeys for the wounded and of grief for the dead as to resolve never to fire at one of the race again.

James Forbes, in his "Oriental Memoirs," relates a striking instance of this kind. One of a shooting party had killed a female monkey in a banian tree, and carried it to his tent. Forty or fifty of the tribe soon gathered around the tent, chattering furiously and threatening an attack, from which they were only diverted by the display of the fowling-piece, whose effects they seemed perfectly to understand. But while the others retreated, the leader of the troop stood his ground, continuing his threatening chatter. Finding this of no avail, he came to the door of the tent, moaning sadly, and by his gestures seeming to beg for the dead body. When it was given, he took it sorrowfully up in his [76] arms and carried it away to the waiting troop. That hunter never shot a monkey again.

This deep feeling for the dead is probably not common among monkeys. The gibbon, for instance, is said to take no notice of the dead. It is, however, highly sympathetic to injured and sick companions, and this feeling seems common to all the apes. No human being could show more tender care of wounded or helpless companions than has often been seen in members of this affectionate tribe of animals.

Without giving further examples of the intelligence and sympathy of the apes, we may say that they possess in a marked degree the mental powers to which man owes so much, viz. observation and imitation. The ape is the most curious of the lower animals—that is, it possesses the faculty of observation in an unusual degree. What we call curiosity in the ape is the basic form of the characteristic which we call attention or observation in man. Its seeming great activity in the ape is what might naturally be expected in an observant animal when removed from its natural habitat to a location where all around it is new and strange. Man under like circumstances is as curious as the ape, while the latter in its native trees probably finds little to excite its special attention. In both man and the ape it needs novelty to excite curiosity.

Again, the ape is imitative in a high degree. This faculty also it does not share with the lower [77] animals, but does with man, imitation being one of the methods by which he has attained his supremacy. Observation, imitation, education, are the three levers in the development of the human intellect. The first two of these the ape possesses in a marked degree. It is susceptible also to the last, being very teachable. Education certainly exists to some extent among the apes in their natural habitat, perhaps to as great an extent as it did in primitive man. In the latter case it is doubtful if there was much that could be called designed education, the young gaining their degree of knowledge by observing and imitating their elders. The same is certainly the case among the apes.

We may reasonably ask what there is in the life and character of the apes to give them this mental superiority over the remaining lower animals. It is certainly not due to the arboreal life and powers of grasp of these animals, for in those respects they resemble the lemurs, which are greatly lacking in intelligence. Whether the monkeys emerged from the lemurs or the two groups developed side by side is a question as yet unsettled; at all events they are closely similar in conditions of existence. Yet while the monkeys are the most intelligent and teachable of animals, the lemurs are among the least intelligent of the mammalia. There is here a marked distinction which is evidently not due to difference of structure or habitat, and must [78] have its origin in some other characteristic, such as difference in life habits.

There is certainly nothing in the diet of the ape to develop intelligence. The frugivorous and herbivorous animals do not need cunning and shrewdness to anything like the extent necessary in carnivorous animals. They do not need to pursue or lie in wait for prey; and they escape from their enemies mainly through strength, speed, concealment, or other physical powers or methods. Escape may occasionally develop mental alertness, but does not usually do so. Certainly if the alert, watchful, suspicious habits of the apes are due to the requisite of avoiding dangerous enemies, we might naturally look for similar habits in the lemurs, which are similarly situated. And if we consider the wide distribution of the apes throughout the tropics of both hemispheres, and their great diversity in species and condition, it seems very unlikely that in all these localities their relations with other animals would be such as to develop the mental alertness which they so generally display. The fact appears to be that, while this may be a cause, it is not a leading cause, of mental development in animals, and that we must seek elsewhere for the origin of animal intelligence.

Research, indeed, leads us to examples of intelligence where we should least expect to find it. Among the mammalia we perceive one marked example in the beavers, the only one in the great [79] class of the rodents, with their nine hundred or more of species. But we must go still lower, to the insects, for the most striking examples, finding them alone in the ants, the bees, and the termites, among the vast multitude of insect forms. Less marked instances appear in the elephants, in some of the birds, and in certain other gregarious animals.

From these examples, and what is elsewhere known of animal intelligence, one broad conclusion may be drawn, that all the strikingly intelligent animals are strongly social in their habits, and that no decided display of intelligence is to be found among solitary species. This conclusion becomes almost a demonstration in the case of the ants and bees. The ants, for instance, comprise hundreds of species, spread over most of the world, mainly social, but occasionally solitary. The social species, while varying greatly in habit, all display powers of intelligence, and these so diversified as to indicate many separate lines of evolution. The solitary ants, on the contrary, manifest no special intelligence, and do not rise above the general insect level. The same may be said of the bees. The hive bee, the most communal in habit, shows the highest traits of intelligent activity. The bees which form smaller groups and the social wasps stand at a lower level, and the solitary bees and wasps sink to the ordinary insect plane. We arrive at like conclusions from observation of the social termites, or white ants, some [80] species of which are remarkable for their intelligent coöperation and division of duties.

Examples similar in kind may be drawn from the vertebrates. Among the birds there are none more quick-witted than the social crows, none with less display of intelligence than the solitary carnivorous species. Birds are rather gregarious than social. There are few species whose association is above that of mere aggregation in flight. Those more distinctively social usually have special habits which indicate intelligence—as in the often cited instances of their seemingly trying and executing delinquents. Among the carnivorous mammals the social dog or wolf tribe displays the intelligent habit of mutual aid. The horses, oxen, deer, and other gregarious hoofed animals have a degree of division of duties, but their intelligence is of a lower grade than that of the dogs and the elephants. On the whole, it may be affirmed that the social habit is frequently accompanied by instances of special intelligence to which we find no counterpart among the solitary forms, and that the highest manifestations of intelligence in the lower animals are found in those forms which possess communal habits, as the ants, bees, termites, and beavers.

One important characteristic of the communal animals is that they become mentally specialized. They round up their powers, build barriers of habit over which they cannot pass, perform the [81] same acts with such interminable iteration that what began as intellect sinks back into instinct. Each individual has fixed duties and is confined within a limited circle of acts, whose scope it cannot pass, or only to the minutest extent.

The non-communal social animals, on the contrary, are not thus restricted. Their intelligence is of a generalized character, and is capable of developing in new channels. None are tied down to special duties, each possesses the full powers of all, and they are thus more open to a continued growth of the intellect than the communal forms. To this class belongs the ape. Its intelligence is general, not special; broadly capable of development, not narrowed and bound in by the limitation of certain fixed and special duties.

The suggestions above offered point to three grades of community among animals, which may be designated the communal, the social, and the solitary. Among these there are, of course, many stages of transition from one to the other. The specially communal, including the ants, bees, termites, and beavers, are those in which there is almost a total loss of individuality, each member working for the good of the community as a unit, not for its personal advantage. The result consists in organized industries, division and specialization of duties, a common home, food stock, etc. At a lower level in animal life, that of the hydroid polyps, communism has become so complete that [82] the community has grown into an actual individual, the members not being free, but acting as organs of an aggregate mass, in which each performs some special duty for the good of the community.

The social animals differ from the communal in that the individuality of the members is fully preserved. There is some measure of work for the group, some degree of mutual aid, some evidence of leadership and subordination, but these are confined to a few exigencies of life, while in most of the details of existence each member of the group acts for itself. The solitary animals are those which do not form groups larger than that of the family, and into whose life the principle of mutual aid, outside the immediate family relations, does not enter. Each acts for itself alone, and intercourse between the individuals of the species is greatly restricted.

The advantages of social habits among animals are evident. There is excellent reason to believe that all animals, and especially such advanced forms as the vertebrates and the higher arthropods, have some power of mental development, some facility in devising new methods of action to meet new situations. Though their reasoning power may be small, it is not quite lacking, and many examples of the exercise of the faculty of thought could be cited if necessary.

What we are here concerned with, is the final result of such exercises of individual thought [83] powers. In the case of the solitary forms, such new conceptions die with the individual. Though they may exert an influence on the development of the nervous system, and aid in the hereditary transmission of more active brain powers, they are lost as special ideas, fail to be taken up and repeated by other members of the species. This is not the case with the social animals. Each of these has some faculty of observation and some tendency to imitation, and useful steps of advance made by individuals are likely to be observed and retained as general habits of the community. Anything of importance that is gained may be preserved by educative influences. The facility of mental communication between these creatures is perhaps much greater than is generally supposed, and acts of importance which are not directly observed might in many cases be transmitted through repetition for the benefit of the group. We know this to be the main agency in human progress. New ideas are of rare occurrence with man. Ideas of permanent value do not occur to one per cent., perhaps not to one hundredth of one per cent., of civilized mankind, yet few of such ideas are lost, and that which has proved of advantage to an individual soon becomes the common possession of a community.

Among the lower animals new and advantageous ideas are probably of exceedingly rare occurrence. When they do occur, their advantage to solitary [84] forms is very slight, being that of minute steps of brain development and hereditary transmission of the same. To social forms they are doubly advantageous, since, while they tend to brain development, they may also be preserved in their original form, and transmitted directly to members of the group. They are still more advantageous to the communal animals, from the closer intercourse of these, and their constant association in acts of mutual aid. But in the latter instance their influence is usually exerted for the benefit of the community as a unit, while in the case of social animals it is of advantage to the individual.

The result of such a process of evolution in the case of the communal animals is a strict specialism. A series of acts of advantage to the community are slowly developed, and are repeated so frequently that they become instinctive, while a fixed circle of duties arises, through whose links it is almost impossible to break. There is no reason to believe that the individual initiative is wanting. The varied round of duties of a community of ants, for instance, could only have arisen through step after step of progress from the condition of the solitary ants. If such steps have been made, others may be made, and are likely to be preserved if found advantageous. The ant individual preserves its powers of observation and thought and may initiate new processes. But most of the ant communities are already so [85] excellently adapted to the conditions of their life as to leave little opportunity for improvement, so that the adoption of new and advantageous habits are certain to be exceedingly rare.

It is an interesting fact that communalism has been confined to animals of comparatively low organization. The most complete examples of it exist in the polyps and some other low forms, in which each community has become a compound individual, the members remaining attached to the parent stock. The next higher examples to be met are the frequently cited ants and bees, belonging to the lowly organized class of arthropoda, yet, through the advantage of association and mutual aid, developing actions and habits only found elsewhere in the human race. The only example among vertebrates is that of the beavers, members of the low order of rodents. With these the results are less varied and intricate than with the ants, in accordance with the much smaller size of the community. All the higher vertebrates are either social or solitary in habit, and among them the narrow specialism of the communal forms does not exist. Each individual works in large measure for itself, its mental powers remain generalized, and it is not tied down to the performance of a series of fixed hereditary acts from which escape is well-nigh impossible.

Of the social animals, man presents the most complete type, and the one from which we can [86] best deduce the conditions of the class. A human community is made up of individuals of many degrees of intellectual ability, the mass remaining at a low level, the few attaining a high level. Yet those of high powers of intellect set the standard for the whole, teach the lower either by precept or example, and aid effectively in advancing the standard of the community. A rope or chain is said to be as weak as its weakest part. A human community, on the contrary, may be said to be as strong as its strongest part. The standing of the whole is dependent upon the thoughts and acts of the few, from whom the general mass receive new ideas and gain new habits. The existing intellectual and industrial position of mankind is very largely a result of ideas evolved by individuals age after age, and preserved as the mental property of the whole. Destroy the books and works of art and industry of any community, cut off its intellectual leaders, remove from the general mind the results of education, and it would at once fall back to a low level and be obliged to begin again its slow climb upward. The intellectual standing of any civilized nation depends upon two things: the preservation in books, in memory, and in works of art and industry, of the ideas of ancient workers and thinkers; and the mental activity of living thinkers and inventors, whose work takes its start from this standpoint of stored-up thought. Rob any community of all its basic ideas, and it [87] would quickly retrograde to a primitive condition of thought and organization, from which it might need many centuries to emerge.

It has been said above that man is the highest example of the social animal. While that is the truth, it is not the whole truth. He is at the same time the highest example of the communal animal. Mutual aid, organization into strictly rounded communities, labor for the good of the whole, is as declared in him as in the most developed community of the ants, and we admire the work of the latter simply because they repeat at a lower level the work of man. In truth, in man we have a splendid example of the existence of the individual initiative in connection with the communal organization. Specialism exists in a hundred forms. Some nations have been tied down by it to conditions almost as fixed as those of the ants. But generalism exists in as full a measure, new ideas are constantly modifying or replacing the old, and the communism of man is a progressive one, steadily borne upward on the wings of new ideas. Individual thought has the fullest swing, and it is to the system of special reward for useful thought and act that man owes much of his great advance. On the other hand, reward without useful service has been one of the leading agencies that have acted to check human progress.

The lower animals do not possess the advantage of man in his power of preserving the thoughts and [88] products of the past as a foundation for new steps of progress. Memory may aid them to a slight degree, but they have no special means of recording useful ideas. This cannot fairly be said of the communal forms, which possess the result of the labors of former generations as useful object lessons. But in the higher animals no means exist for the permanent preservation of ideas, and each step of progress must be due to the direct influence of living individuals and the indirect result of natural selection.

This is one cause of the slow mental advance of the lower animals. A second is the deficiency in educational influences, which have had so much to do with human progress. Education is not quite wanting in the brute creation. There are many instances on record of instruction given by the adults to the young. But this agency is in its embryo stage, and its influence must be small. Again, each tribe of lower animals is apt to fall into a fixed circle of life acts, to become so closely adapted to some situation or condition that any change of habits would be likely to prove detrimental. This is a state of affairs tending to produce stagnation and vigorously to check advance. Many instances of this could be cited from human history, while it is the common condition with the animals below man.

To return to the apes, the considerations above taken lead to the conclusion that it is chiefly, if not [89] solely, to their social habits that they owe their mental quickness. While only in minor traits communal, they are eminently social, and have doubtless derived great advantage from this. The lemurs, which share their habitat and resemble them in organization, are markedly unsocial, and are as mentally dull as the apes are mentally quick. Possibly, the thought powers of the apes once set in train, there may have been something in the exigencies of arboreal life that quickened their powers of observation; but we are constrained to believe that the main influence to which they owe their development is that of social habits, in which they stand at a high, if not the highest, level among the distinctly social animals.

The thought capacities of the ape intellect are general, not special. The mind of these animals remains free and capable of new thought in new situations. It is fully alive to the needs and dangers of arboreal life, and advances no farther in its native habitat because there is nothing more of importance to be learned. But while fixed it is not stagnant. When the ape is taken from its native woods and put among the many new conditions arising on shipboard and in human habitations, we quickly perceive indications of its mental alertness. Its faculties of observation and imitation are actively exercised, and new habits and conceptions are quickly gained. Could the apes be made to breed freely in captivity, so that a domestic race, [90] comparable to that of the dogs, could be obtained, their mental powers might, perhaps, be cultivated to an extraordinary degree, yielding instances of thought approaching that of man. The ape is especially notable for its tendency to attempt new acts of itself, not waiting to be taught, as in the case of other domesticated animals. In short, it seems by all odds to be the animal best fitted mentally to serve as the basis of a high intellectual development, as it is the best fitted physically to change from the attitude of the quadruped to that of the biped.

The anthropoid apes in general manifest a reversion from the social toward the solitary state, this condition reaching its ultimate in the orang, which is one of the most solitary of animals. The smaller forms are the most social, the gibbons being decidedly so. There is very good reason to believe that the man-ape was highly social, if we may judge from what we find in all races of men, and all grades, from the savage to the civilized. This animal was thus in a position to avail itself of all the advantages of the social habit, and to gain the mental development thence arising. How long ago it was when it left the trees and made its home upon the ground, it is impossible to say. It may have been as far back as the early Pliocene or the late Miocene Period, or even earlier. As yet its brain was probably no more developed than in the case of the other anthro [91] poids, perhaps less so than in the existing species. But in its new habitat it was exposed to a series of novel conditions that must have exerted a healthful and stimulating influence upon its mind.

If it had remained in the trees we should probably to-day have only a man-ape still. Leaving their safe shelter for the ground, it became exposed to new dangers and was forced to fit itself to fresh conditions. Prowling carnivorous animals haunted its new place of residence, and these it had to avoid by speed or alertness of motion, or combat them by strength and the use of weapons. The carnivorous tastes which it had in all probability gained, made it a creature of the chase, pursuing swift animals, capturing them by fleetness or stratagem, or bringing them down with the aid of clubs and missiles. Such a new series of duties and dangers could not fail to exert a vigorous influence upon a brain already quick of thought and susceptible to fresh impressions, and we may well conceive that the man-ape then entered upon a new and rapid phase of mental progress, its brain developing in powers and growing in dimensions as it slowly became adapted to its new situation and grew able to cope with fresh demands and critical exigencies.

There is still another influence which has had its share, perhaps a very prominent share, in the intellectual development of animals, yet which no writer seems to have considered from this point of [92] view. The probable effect of this influence needs to be taken into account, in conclusion of this section of our subject. It is that of the comparative agency of the senses in the development of the mind, and the effects likely to arise from the dominance of some one of the senses.

In the lowest animals touch was the predominant, if not the only sense, taste perhaps being associated with it. But these senses, which demand actual contact with objects, obviously could give none but the narrowest conception of the conditions of nature. The other senses, sight, hearing, and smell, give intimations of the existence and conditions of more or less distant objects, and their development greatly widened the scope of outreach in animals and must have exerted a powerful influence upon the growth of mental conditions.

It need scarcely be said that the sense which gives the fullest and most extended information about existing things is necessarily the one that acts most effectively upon the mind, and that this sense is that of sight. Hearing and smell yield us information concerning certain local conditions of objects, but sight extends to the limits of the universe, while in regard to near objects it has the advantage of being practically instantaneous in action and much fuller in the information it conveys. Sight, therefore, is evidently the most important of the senses, so far as the broadening of the mental [93] powers is concerned, and any animal in which it is predominant must possess a great advantage in this respect over those species controlled to any great degree by one of the lower senses.

It may be said here that sight only slowly gained dominance in animal life. Though the eye, as an organ of vision, is found at a low level in the animate scale, the indications are that it long played a subordinate part, and has gained its full prominence only in man. During long ages life was confined to the sea, hosts of beings dwelling in the semi-obscurity of the under waters, and great numbers at too great a depth for light to reach them. To vast multitudes of these sight was partly or completely useless. The same may be said of hearing, the under-water habitat being nearly or completely a soundless one. The only one of the higher senses likely to be of general use to these oceanic forms is that of smell, and it may be that their knowledge of distant objects was mainly gained through sensitiveness to odors.

Of invertebrate land animals the same must be said. The land mollusks and the great order of insects and other land arthropods only to a minor extent dwell in the open light. Very many species haunt the semi-obscurity of trees or groves, hide among the grasses, lurk under bark, sticks, and stones, or dwell through most of their lives underground. Hosts of others are nocturnal. To only a small percentage of insects can sight be of any [94] great utility, while hearing seems also to be of slight importance. Smell is probably the principal sense through which these animals gain information of distant objects.

There is existing evidence that the sense of smell in some insects is remarkably acute. The imprisoned female of certain nocturnal species, for instance, will attract the males from a comparatively immense distance, under conditions in which neither sight nor hearing could have been brought into play. The emission of odors and acute sensibility to them is the only presumable agency at work in those instances. As regards the most intelligent of the insects, the ants and the termites, the former are largely subterranean, the latter not only subterranean, but blind. In the one case, sight can play only a minor part, in the other, it plays no part at all. Touch and smell seem to be the dominant senses in these animals, and the degree of intelligence they display shows of how high a development these senses are susceptible. Yet the intelligence arising from them must necessarily be local and limited in its application; it cannot yield the breadth of information and degree of mental development possible under the dominance of sight.

In the vertebrates we find a fully developed and broadly capable organ of vision, and it might be hastily assumed that in those animals sight is the dominant sense. But there are numerous facts [95] which lead to a different conclusion. Many of the vertebrates are nocturnal, many dwell in obscure situations, many in the total darkness of caverns, underground tunnels and excavations, or the ocean's depths. To all these sight must be of secondary importance. Hearing also can be of no superior value, and the dominant sense must be that of smell. In the bats there would appear to be a remarkably acute power of touch, if we may judge from the facility with which they can avoid obstacles at full flight after their eyes have been removed.

It might, however, be supposed that in the higher land vertebrates sight is predominant, and that the diurnal mammals depend principally upon their eyes for their knowledge of nature. But there are facts which throw doubt upon this supposition. These facts are of two kinds, external and internal. That the quadrupeds, in general, are highly sensitive to odors is well known, and also that they trust very largely to the sense of smell. Hunters are abundantly aware of this, and have to be quite as careful to avoid being smelt by their game as to avoid being seen. We have abundant evidence of the remarkable acuteness of this sense in so high an animal as the dog, which can follow its prey for miles by scent alone, and can distinguish the odors, not only of different species, but of different individuals, being capable of following the trail of one person amid the tracks of numerous others.

[96] The internal evidence of this fact is equally significant. In the vertebrates, in general, the olfactory lobe of the brain is largely developed, much exceeding in size the lobe of the optic nerve. It forms the anterior portion of the cerebrum, and in many instances constitutes a large section of that organ, being marked off from it by only a slight surface depression. If we can fairly judge, then, by anatomical evidence, the sense of smell plays a very prominent part in the life of all the lower vertebrates. If we take our domestic animals as an example, the olfactory lobe of the horse is considerably larger

than that of man, though the brain, as a whole, is very much smaller, so that, comparatively, this organ constitutes a much larger portion of the total brain. The other domestic animals yield similar evidence of the great activity of the sense of smell.

While there is no doubt that sight is an active sense in all the higher quadrupeds, it evidently divides this activity with smell to a much greater degree than is the case with man, in whom smell plays a minor part, sight a major part, among the organs of sense.

This fact shows its effect in the comparative mental development of man and the lower animals. Man, depending so largely on vision, gains the broadest conception of the conditions of nature, with a consequent great expansion of the intellect. The quadrupeds, depending to a considerable [97] degree upon smell for their conceptions of nature, are much narrower in their range of information and lower in their mental development. As regards the ape family, it occupies a position between man and the quadrupeds, and its intellectual activity may well be due in great measure to an increased trust in sight and a decreased trust in smell in gaining its conception of nature.

The question may arise, Why, if sight has this superiority over smell, did it not long since gain predominance, and relegate smell to a minor position? It may be answered that the superiority of sight is not complete. In one particular this sense is inferior to smell. The leading agency in the development of the sense organs of animals has been the struggle for existence, including escape from enemies, and the perception of food-animals or material. In these processes acuteness of smell plays a very important part. It has, moreover, the advantage of gathering information from all directions, while sight is very limited in its range. The eye is so subject to injury that its multiplication over the body would be rather disadvantageous than otherwise, while, localized as it is, a movement of the head is necessary to any breadth of vision, and the whole body must rotate to bring the complete horizon under observation. It seems evident, from these considerations, that sight is much inferior to smell in the timely perception of many forms of danger. Light comes in straight lines [98] only, and a movement of the body is necessary to perceive perils lying outside these lines. Odors, on the contrary, spread in all directions, and make themselves manifest from the rear as well as the front.

In all probability this fact has had much to do with the continued dependence of animals on smell. In fishes and reptiles a full sweep of vision is so slowly gained that some more active sentinel sense is requisite to safety. In mammals the head rotates more easily, but valuable time is lost in the rotation of the whole body. These animals, therefore, depend on both sight and smell, in some cases equally, in some more fully on one or the other of these senses. When we reach the semi-upright ape, we have to do with a form capable of turning the body and observing the whole surrounding circle of objects more quickly and readily than any quadruped. As a result, these animals have grown to depend more fully on vision and less on smell than the quadrupeds. Finally, in fully erect man, the power of quick turning and alert observation of the whole circle of the horizon reaches its ultimate, and in man sight has become in a large degree the dominant sense, and smell has fallen to a minor place.

With this change in the relations of the senses has come a change in the degree of mental development. It is highly probable that the dependence [99] of the apes on vision instead of smell has had much to do with their mental activity, quickness of observation, and active curiosity. In man there can be no question that it has played a great part in the rapid development of his intellectual powers, and in the extraordinary breadth of his conception of nature as compared with that of the lower animals. While hearing and smell advise us of neighboring conditions only, and have their chief utility as aids to the preservation of existence, sight makes us aware of the conditions of nature in remote localities, extending far beyond the limits of the earth. While this sense plays its part as one of the protective agencies, it is still more useful as an agent in the acquisition of knowledge in general, and has much to do with the development of the intellectual faculties. We may look, therefore, upon the increasing dominance of the sense of sight as a leading agency in the making of man as a thinking being, and may ascribe to this in a considerable measure the thirst for information and faculty of imitation so marked in the apes.

[100]

VII

THE ORIGIN OF LANGUAGE

One of the characteristics of man, of which we spoke as among those to which his high development is due, is that of language. There is nothing that has had more to do with the mental progress of the human race than facility in the communication of thought, and in this vocal language is the principal agent and in the fullest measure is the instrument of the mind. Human speech has, in these modern times, become remarkably expressive, indicating all the conditions, relations, and qualities, not only of things, but of thoughts and ideal conceptions. And the utility of language has been enormously augmented by the development of the arts of writing and printing. Originally thought could only be communicated by word of mouth and transmitted by the aid of the memory. Now it can be recorded and kept indefinitely, so that no useful thought of able thinkers need be lost, but every valuable idea can be retained as an educative influence through unnumbered ages.

In this instrumentality, which has been of such extraordinary value to man, the lower animals are strikingly deficient. They are not quite devoid of [101] vocal language, though it is doubtful if any of the sounds made by them have a much higher linguistic office than that of the interjection. But emotional sounds, to which these belong, are not destitute of value in conveying intelligence. They embrace cries of warning, appeals to affection, demands for help, calls for food supplies, threats, and other indications of passion, fear, or feeling. And the significance of these vocal sounds to animals may often be higher than we suppose. That is, they may not be limited to the vague character of the interjection, but may occasionally convey a specific meaning, indicative of some object or some action. In other words, they may advance from the interjection toward the noun or the verb, and approach in value the verbal root, a sound which embraces a complete proposition. Thus a cry of warning may be so modulated as to indicate to the hearer, "Beware, a lion is coming!" or to convey some other specific warning. We know that accent or tone plays a great part in Chinese speech, the most primitive of existing forms, a variation in tone quite changing the meaning of words. The same may be the case with the sounds uttered by animals to a much greater extent than we suppose.

We know this to be the case with some of the birds. The common fowl of our poultry yards has a variety of distinct calls, each understood by its mates, while special modulations of some call or cry are not uncommon among birds. The mam [102] malia are not fluent in vocal powers, their range of tones being limited, yet they certainly convey definite information to one another. Recent observers have come to the conclusion that the apes do, to a certain extent, talk with one another. The experiments to prove this have not been very satisfactory, yet they seem to indicate that the woodland cries of the apes possess a certain range of definite meaning.

We are utterly ignorant of what powers of speech the man-ape possessed. It must, in its developed state as a land-dwelling, wandering, and hunting biped, have needed a wider range of utterance than during its arboreal residence. It was exposed to new dangers, new exigencies of life affected it, and its old cries very probably gained new meanings, or new cries were developed to meet new perils or conditions. In this way a few root words may have been gained, rising above the value of the interjection, and expressing some degree of definite meaning, though still at the bottom of the scale of language, the first stepping stones from the vague cry toward the significant word.

Between this stage and that of human language an immense gap supervenes, a broad abyss which it seems at first sight impossible to bridge. As the facts stand, however, it has been largely bridged by man himself. Side by side with the highly intricate languages which now exist, are various [103] primitive forms of speech which take us far back toward the origin of human language. So advanced a people as the Chinese speak a language practically composed of root words, the higher forms of expression being attained by simple devices in the combination of these primitive word forms. The same may be said, in a measure, of ancient Egyptian speech. We can conceive of an early state of affairs in which these devices of word compounding were not yet employed, and in which each word existed as a separate expression, unmodified by association with any other word. Among the savage races of the earth very crude forms of language often exist, the methods of associating words into sentences being of the simplest character, though few surpass the Chinese in simplicity of system.

But all this represents an advanced stage of language evolution, a development of thought and its instrument which has taken thousands of years to complete. We cannot fairly judge from it what the speech of primitive man may have been, for in every case there has been a long process of development; aided, no doubt, in many cases, by educative influences acting from the more advanced upon the speech of the less advanced races.

If we seek to analyze any of these languages, the most intricate as well as the least advanced, we find ourselves in most instances able to isolate the root word as the basic element of speech. [104] From this simple form all the more developed forms seem to have arisen. Take away their combining devices, and the root words fall apart like so many beads of speech, each with a defined significance of its own and fully capable of existing by itself. The Aryan and the Chinese especially offer themselves to this analytic method. Strip off the suffixes and affixes from Aryan words, get down to the germinal forms from which these words have grown, isolate these germs of speech, and we find ourselves in a language of root forms, each of which has grown vague and wide in significance as the modifying elements that limited its meaning have been removed. In the Chinese the problem is a much simpler one. We need simply to take the existing words out of their place in the sentence and let them stand alone, and we have root words at first hand. We may go through the whole range of human speech and, with more or less difficulty, arrive at a similar result. In short, the evidence seems conclusive that the language of mankind began in the use of isolated words of vague and broad significance, and that all the subsequent development of language consisted in the combination of these words, with a modification and limitation of their meaning, the families of speech differing principally in the method of combination devised.

It must, indeed, be said that in isolating the root forms of modern languages we reach conditions still [105] far removed from those of primitive speech. These roots are in a measure packed with meaning. Time has added to their significance, and they lack the simplicity they probably once possessed. In particular, they have gained ideal senses, entered in a measure into that broad language of the mind which has been gradually added to the language of outer nature. The recognition of the existence of mind and thought doubtless came somewhat late in human development. Man long knew only his body and the world that surrounded it. Step by step only did he discover his mind. And when it became necessary to speak of mental conditions, no new language was invented, but old words were broadened to cover the new conditions. The mind is analogous to the body in its operations, ideas are analogues of things, and it was usually necessary only to add to the physical significance of words the corresponding ideal significance. In this way a secondary language slowly grew up, underlying and subtending the primary language, until the words invented to express the world of things were employed to include as vast a world of thoughts.

In getting down, then, to the language of primitive man we are obliged to divest the root forms of speech of all this ideal significance, and confine them to their physical meanings. In dealing with the languages of the least advanced existing tribes of mankind, indeed, little of this is requisite. The [106] language of the mind with them has not yet begun its growth or is in its first simple stages. Only half the work of the evolution of language is completed. There is, indeed, no tribe so undeveloped as to use the primitive forms of speech. The most savage of the races of mankind have made some progress in the art of combining words, gained some ideas of syntax and grammatical forms. Yet in certain instances the progress has been very slight, and in all we can see the living traces of the earlier method of speech from which they emerged.

It is to the ability to think abstractly and to form words with an abstract significance that human language owes much of its high development. But this ability is largely confined to civilized mankind, savages being greatly or wholly lacking in it. This deficiency is indicated in their modes of speech. Thus a native of the Society Islands, while able to say "dog's tail," "sheep's tail," etc., has no separate word for tail. He cannot abstract the general term from its immediate relations. In the same way the uncivilized Malay has twenty different words to express striking with various objects, as with thick or thin wood, a club, the fist, the palm, etc., but he has no word for "striking" as an isolated thought. We find the same deficiency in the speech of the American Indians. A Cherokee, for instance, has no word for "washing," but can express the different kinds of washing by no less than thirteen distinct words.

[107] All this indicates a primitive stage in the evolution of language, one in which every word had its immediate and local application, while in each word a whole story was told. The power of dividing thought into its separate elements was not yet possessed. As thought progressed men got from the idea of "dog" to that of "dog's tail." They could not think of the part without the whole. Then they reached a word for "dog's tail wags." But the idea of "wags" as an abstract motion was beyond their powers of thought. They could not think of action, but only of some object in action. The language of the American Indians was an immediate derivation from this mode of word formation, every proposition, however intricate it might be, constituting a single word, whose component parts could not be used separately. The mode of speech here indicated is one form of development of the root. Other forms are the compounding of the Chinese and the Mongolian and the inflection of the Aryan and the Semitic, all pointing directly back to the root form as their unit of growth.

The inference to be drawn from all this is that the language of primitive man consisted of isolated words, sounds which may originally have been mere cries or calls, but which gradually gained some definiteness of meaning, as signifying some of the varied conditions of the outer world. This is the conclusion to which philologists have now very generally come. The recognition that language [108] consists of root words, variously modified and combined, leads back irresistibly to a period in which those roots had not yet begun to be modified and combined. The roots are the hard, persistent things in human speech. Grammatical expedients are the net in which these roots have been caught and confined. Free them from the net, and it falls to pieces, while the roots remain intact, the solid and persistent primitive germs of speech.

Yet in isolating root language as the basis of grammatical language we go far toward closing the gap between animal and human speech. It is still, doubtless, of considerable width, yet the distinction is no longer one of kind, but is simply one of degree. Primitive man had a much greater scope of language than is possessed by any of the lower animals, and the vocal sounds used had a clearer and more definite significance; but their nature was the same. They doubtless began in calls and cries like those in use by animals, and though these had increased in number and gained more distinct meanings, the difference in character was not great. In short, the analytic method employed by modern philologists has gone far to remove the supposed vast distinction between brute and human speech, and has traced back the language of man to a stage in which it is nearly related in character to the language of animals. The distinction has been brought down to one of degree, scarcely one of kind. A direct and simple process of evolution [109] was alone needed to produce it, and through that evolution man undoubtedly passed in his progress upward from his ancestral stage.

The language of the lower animals is a vowel form of speech. It lacks the consonantal elements, the characteristic of articulation. In this man seems to have at first agreed with them. The infant begins its vocal utterances with simple cries; only at a later age does it begin to articulate. If we may judge from the development of language in the child, man began to speak with the use of sounds native to the vocal organs, and progressed by a process of imitation, endeavoring to reproduce the sounds heard around him: the voices of animals, the sounds of nature, etc. This tendency to imitate is not peculiar to man. It exists in many birds, and in some attains a marked development. The mocking bird, for instance, has an extraordinary flexibility of the vocal organs and power of imitating the voices of other birds. The parrot and some other birds go farther in this direction, being capable of using articulate language and clearly repeating words used by man.

None of the mammalia possess this facility. It is not found in the apes, and probably was not possessed by the ancestor of man. But it is not difficult to believe that in the efforts of the latter to gain a greater variety of vocal utterance, its organs of speech became more flexible, and in time it gained the power of articulation.

[110] There are races of existing men whose powers of language seem still in the transition stage between articulate and inarticulate

speech. This seems the case with the Bushmen and Hottentots of South Africa, whose vocal utterances consist largely of a series of peculiar clicks that are certainly not articulate speech, though on the road toward it. The Pygmies of the Central African forests seem similarly to occupy an intermediate position in the development of language. Those who have endeavored to talk with them speak of their utterance as being inarticulate in sound. It appears to be a sort of link between articulate and inarticulate speech. In short, the great abyss which was of old thought to lie between the languages of man and the lower animals has largely vanished through the labors of philologists, and we can trace stepping-stones over every portion of the wide gap. The language of man has not alone been evidently a product of evolution, but also one of development from the vocal utterances of the lower animals; and the man-ape, in its slow and long progress from brute into man, seems to have gradually developed that noble instrument of articulate speech which has had so much to do with subsequent human progress.

[111]

VIII

HOW THE CHASM WAS BRIDGED

In his bodily formation the man-ape differed little from man. The differences which existed were probably of a minor character, no greater than could readily exist within the limits of a species. If this assertion be questioned, it seems sufficient to call attention to the recent researches into the anatomy of the anthropoid apes, which differ in species, if not in genera, from man, yet are closely similar to him in all their main features of organization. Even in the brain, to whose great development man owes his superiority, the only marked difference is in size. Structurally, the distinctions are unimportant. If, then, these distant relatives so closely resemble man in physical frame, his immediate relative in the line of descent must have approached him still more closely in organization. After this ancestor had become a true, surface-dwelling biped, the differences in structure were probably so slight that physically the two forms were in effect identical. The man-ape was, as there is reason to believe, considerably smaller than man, perhaps about equal in size and stature to the chimpanzee, but [112] that does not constitute a specific difference. There may have been some differences in the skeletal and muscular structure. The vocal organs, for instance, probably differed, the evolution of language in man being accompanied with certain changes in the larynx. The skull was certainly much more ape-like. Yet variations of this kind, due to differences in mode of life, are minor in importance, and may easily come within the limits of a species. While the great features of organization remain intact, small changes, due to new exigencies of life, may take place without affecting the zoölogical position of an animal. The most striking difference between man-ape and man, that of the development of the brain to two or three times its size and weight, is similarly unessential in classification while the brain remains unchanged in structure. That it has remained unchanged we may safely deduce from the close similarity between the brain of man and those of the existing anthropoid apes. The cause of the increase in size is so evident that it need only be referred to. Since the era of the man-ape, almost the whole sum of the forces of development have been centred in the mental powers of this animal, with the result that the brain has grown in size and functional capacity, while the remainder of the body has remained practically unchanged.

That man as an animal has descended from the lower life realm, none who are familiar with the [113] facts of science now think of denying. This has attained to the scientist, and to many non-scientists, the level of a self-evident proposition. But that man as a thinking being has descended from the lower animals is a different matter, concerning which opinion is by no means in unison. Even among scientists some degree of difference of opinion exists, and such a radical evolutionist as Alfred Russell Wallace finds here a yawning gap in the line of descent, and is inclined to look upon the intellect of man as a direct gift from the realm of spirits. His explanation, it is true, is more difficult than the problem itself. There are no facts to sustain it, and even if he were not able to see how man's mind could be developed by natural selection, it is a sort of *reductio ad absurdum* to call in the angels to bridge the chasm.

Romanes has dealt with the subject from a different and more scientific point of view, and seems to have succeeded in showing that man's intellect at its lowest level is not different in kind from the brute intellect at its highest level. Controversy on this subject is too apt to be based on the difference between the intellect of the brute and that of enlightened man, in disregard of the great mental gap which exists between the latter and the thought powers of the lowest savage. In the preceding section an effort was made to show how crude and imperfect must have been the language of primitive man. Its imperfection was a fair gauge of [114] that of his powers of thought. His intellect stood at a very low level, seemingly no further above that of the highest apes than it was below that of enlightened man.

In fact, enormous as is the interval between the mind of the brute and that of the man of modern civilization, the whole long line of mental development can be traced, with the exception of a comparatively small interval. This is the gap between the intellect of the anthropoid ape and that of primitive man, the one important last chapter in the story of mental evolution. Supernaturalism, driven from its strongholds of the past, has taken its last stand upon this broken link, claiming that here the line of descent fails, and that the gap could not have been filled without a direct inflow of intellect from the world of spirits or an immediate act of creation from the Deity.

This view of the case is not likely to be accepted as final. Science has bridged so many gaps in the kingdom of nature that it is not likely to retire baffled from this one, but will continue its investigations in place of accepting conclusions that have not the standing even of hypothesis, since they are unsupported by a single known fact. At first sight, indeed, the facts which bear upon this question seem stubborn things to explain by the evolution theory. The gap in intellect between the highest apes and the lowest man is a considerable one, which no existing ape seems likely ever to [115] cross. However the anthropoid apes gained their degree of mental ability, it does not appear to be on the increase. They are in a state of mental stagnation and may have remained so for millions of years. Something similar, indeed, can be said of the lowest savages. They also are mentally stagnant. The indications are that for thousands, or tens of thousands, of years in the past their intellectual progress has been almost nothing. Yet it is beyond reasonable question that the advanced thinker of to-day has evolved from an ancestor as low in the mental scale as this savage, probably much lower; and this renders it very conceivable that a similar process of evolution covered the interval between the ape intellect and that of primitive man.

Somewhere, at some time in the far past, the mental stagnation of man was broken, and the development of the mind began its long progression toward enlightenment. This was not in the localities in which the lower savages are now found, the equatorial forests of Africa and South America and other realms of savage life, the change in all probability taking place elsewhere, under new and severe exigencies of life. Similarly we have much justification in saying that somewhere, at some time, the mental stagnation of the ape was broken, and the long development of the mind from ape to man began. This did not take place in the instances of the existing anthropoids, and, as in the [116] analogous case of civilized man, its influencing cause must be looked for in exigencies of existence acting upon some form different in character and habitat from these apes.

The existing anthropoid apes may justly be compared in condition with the existing low savages. In both cases a satisfactory adaptation to their situation has been gained. These apes are still arboreal and frugivorous, as their remote ancestors were. They have for ages been in a state of close adaptation to their life conditions, and the influences of development have been largely wanting. Such evolution as took place must have been extremely slow. In like manner the lowest savages live in intimate relations with the conditions surrounding them. All problems of food-getting, habitation, climate, etc., have long since been solved, and in the tropical forests in which so many of them dwell they are in thorough accord with the situation. Mentally, therefore, they are practically at a standstill and have remained so for thousands of years. The two cases are parallel ones. We can safely say that the later development of man took place in other situations and under other conditions. We may fairly say the same in regard to the ape. Vigorous influences must have been brought to bear upon the ancestor of man as the instigating causes of its mental development into man; and similarly vigorous influences must have been brought to bear upon primitive man to set in [117] train his mental development into intellectual man. And the general character of these influences in both cases may readily be pointed out. An extraordinary development has taken place in the human intellect within a few thousands, or tens of thousands, of years, yielding the difference which exists between the cultivated man of to-day and the debased savage who probably preceded him, and whose counterpart still exists. This has undoubtedly been due to influences of the highest potency. If we can show that influences of equal potency acted upon man's ancestor, we shall have done much toward indicating how the ape brain may have grown into the brain of man.

In both cases the main agency was in all probability that of conflict. Both ape and man, as we take it, developed through some form of warfare. In the former case it was warfare with the animal kingdom; in the latter it was warfare with the conditions of nature and with hostile man. Each of these has been potent in its effects, and to each we owe the completion of a great stage in the evolution of man.

In the tropics, the home of the anthropoid apes of to-day and, probably, of the animal we have named the man-ape, war between man and nature scarcely exists. Nature is not hostile to man. There is no occasion for clothing and little for habitation. Food is abundant for the sparse populations. Little exertion is called for to sustain [118] life. Mental stagnation is very likely to supervene. Yet there, as elsewhere, conflict has had much to do with such mental progress as exists. Mastery in warfare is due to superior mental resources, which gradually arise from the exigencies of conflict, and manifest themselves in greater shrewdness or cunning, superior ability in leadership, better organization, fuller mutual aid, and the invention of more destructive weapons and more efficient tools. War acts vigorously on men's minds, peace acts sluggishly. In the former case man's most valued possession, his life, is in jeopardy, and his utmost powers are exerted for its preservation. Every resource within his power is brought to bear to save himself from wounds or death and to destroy his enemies. If the foes are equal physically, victory is apt to come to those which are superior mentally, which are quicker at devising new expedients, more alert in providing against danger, more skilful in the use of weapons, abler in combining their forces to act in unison. In short, the whole story of mankind tells us that mental evolution has been greatly aided by the influences of warfare, the reaction upon the mind of the effort at self-preservation, the destruction of those at a lower level of intellectual alertness, the preservation of the abler and more energetic, the effect of conflict in bringing into activity all the resources of the intellect, and the hereditary transmission of the powers of mind thus developed. It [119] is, undoubtedly, to war between man and man, and the conflict with the adverse conditions of nature in the colder regions of the earth, that man's development from his lowest to his highest intellectual state has been largely due. This is by no means to say that war is still necessary for this result. Other influences are now at work, of equal or superior potency, and while the conflict with nature and the conditions of society is still of importance, war between man and man is no longer necessary as a mental stimulant. The time was, and that not very far in the past, when it was an essential element in human development.

If we descend to the lowest existing savages, however, it is to find this agency almost non-existent. We can perceive in them no organized warfare and no alert conflict with nature. They are as yet at the very beginning of this stage of evolution, and it certainly exerts little influence upon them. Nature is not adverse, life needs little thought or exertion, they accept the world as they find it, without question or revolt, and their thoughts and habits are as unchangeable as the laws of the Medes and Persians. But the fact that active warfare does not now exist among the lowest tribes of mankind, does not argue that such a state has never existed. In truth, we maintain that primitive man is the outcome of an active and long-continued warfare, and that his settled and sluggish condition today is the ease that follows [120] victory. He has conquered and is at rest after his labors.

For if we compare primitive man with the anthropoid apes, it is to find one striking and important difference between them. The anthropoids are at a level in position with their animal neighbors. Man is lord and master of the animal kingdom, the dominant being in the world of life. He has no rival in this lordship, but stands alone in his relation to the animal kingdom. He is feared and avoided by the largest and strongest beasts of field and forest. He does not fight defensively, but offensively, and whatever his relation to his fellow-man, he admits no equal in the world of life below him. He is the only animal that has made a struggle for lordship. The gorilla is said to attack the lion and drive it from its haunts. If it does so, it is not with any desire for mastery, but simply to rid itself of a dangerous neighbor. The battle for dominion has been confined to man, and in the winning of it no small degree of mental development must have taken place.

The supremacy of man was not gained without a struggle, and that a severe and protracted one. The animal kingdom did not yield readily to man's lordship, and the war must have been long and bitter, settled as the relations now seem. Rest has succeeded victory. The lower animals are now submissive to man, or retire before him in dread of his strength and resources, and the strain upon his [121] powers has ceased. So far as this phase of evolution is concerned the influences aiding the mental development of man have lost their strength. The warfare is over, and man reigns supreme over the kingdom of life.

Of all animals the man-ape was the best adapted for such a struggle. The other anthropoid apes, while favored by the formation of their hands, lacked that freedom of the arms to which man mainly owes his success. No other animal has ever appeared with arms freed from duty in locomotion and at the same time endued with the power of grasping, and these are the features of organization to which the evolution of the human intellect was wholly due in its first stages. The man-ape was not able to contend successfully with the larger animals by aid of its natural weapons. Its diminutive size, its lack of tearing claws, and its lesser powers of speed, left it at a disadvantage, and had it attempted to conquer by the aid of its strength and the seizing and rending powers of teeth and nails, its victory over the larger animals would never have been won. Even with the aid of the cunning and alertness of the apes, their power of observation, their combination for defence and attack, and their general mental superiority to the tenants of the animal world, their supremacy in the event of their becoming carnivorous must have been confined to the smaller creatures, and could not have been established over the larger [122] animals of their native habitat except through the aid of other than their natural powers.

It was by the use of artificial weapons that the conquest was gained. The tendency to use missiles as weapons of offence and defence, which is shown by various species of monkeys, was in all probability greatly developed by the man-ape, the only carnivorous member, if our premises are correct, of the whole extensive family of the apes, and the only one with the free use of its hands and arms. By the use of weapons of this kind the powers of offence of this animal were enormously increased. As skill was acquired in their use, and more efficient weapons were selected or formed, the man-ape steadily advanced in controlling influence, and the lower animal world became more and more subordinated. No doubt the struggle was a protracted one. The previously dominant animals did not submit without a severe and long-continued contest. Thousands of years may have passed before the larger animals were subdued, for it is probable that the invention of superior weapons by an animal of low mental powers was a very slow process. Each stage of invention gave higher success, but these stages were very deliberate ones.

However this be, we can be assured that the superiority of the ancestral man lay in his mental resources, and that his victory was due to the employment of his mind rather than of his body. As a result, the developing influence of the conflict [123] was exerted upon his brain, the organ of the mind, far more than upon his physical frame, and this organ gradually increased in size, while the body as a whole remained practically unchanged. The conflict began with the man-ape on a level in power and dominance with animals of its own size and inferior to those of greater size and strength. It ended with man dominant over all the lower animals. Such a progress, if made by any animal through variation in physical structure, must have caused radical and extraordinary changes in size, strength, and utility of the natural organs of offence. If made, as in the instance in question, through development of the organ of the mind alone, it could pot but have produced a great increase in the size and power of this organ; and the dimensions of the brain in primitive man, as compared with those of the brain in the anthropoid apes, do not seem too great for the magnitude of the result.

The conflict ended, a new animal, man, finally and fully emerged from the family of the apes and settled down in the restful consciousness of victory, with a much larger brain and greatly superior mental powers than were possessed at the beginning of the struggle, yet in physical aspect not greatly changed from his ancestral form after it had first fully gained the erect attitude. The powers gained enabled early man easily to hold the position he had won, and there was no further [124] special strain upon his faculties until a new contest began, that between man and nature, supplemented by a still more vital struggle, that between man and man.

To return to the point from which we set out, it may be said that, as the man-ape gained facility in walking in the erect attitude, and its hands and arms became fully adapted to the use of weapons, its standing in the animal kingdom changed essentially from that before held. Fear and flight ended, retreat ceased, attack began, pursuit succeeded flight, and the great battle for mastery entered upon its long course. An element which aided materially in the victory was the social habit of the animal in question, and the mutual aid which the members of any group gave one another. Educative influences also naturally follow association, every invention or improvement devised by one becomes the property of the whole, and nothing of importance once gained is lost.

The stages of this progress were, undoubtedly, in their outer aspect, stages of improvement in weapons. We seem to see ancestral man, in his early career as a carnivorous animal, seizing the stones and sticks that came readily to hand, and flinging them with some little skill at his prey, in the same manner as we can perceive the baboon doing the same thing. In like manner we observe him breaking off branches from the trees and using them as clubs. One of the first steps of develop [125] ment from this crude stage in the use of weapons would be the selection of stones suited by size and shape for throwing, and the choice of clubs of suitable length and thickness, the latter being stripped of their twigs.

For a long time fresh weapons, those immediately at hand, would be seized and used for every new conflict; but as the idea of the superiority of some weapons to others arose, a second stage of evolution must have begun. The selected club, broken from the tree and prepared for use with some care, and thus embodying a degree of choice and labor, would be too valuable to fling idly away, and might be retained for future use, the first personal possession of inchoate man. Similarly, stones carefully chosen for their suitability for throwing would be probably kept, and a small store of them collected. In short, we may conceive of the man-ape thus gathering a magazine of weapons,—clubs and stones,—sought or shaped during hours of leisure for use in hours of conflict. In this way our animal ancestor doubtless slowly became a skilful hunter, carrying his weapons with him in the chase, and using them efficiently in the conquest of prey.

A third stage in this progress was reached when to some wise-headed old man-ape came the idea of combining the two forms of weapon in use, of fastening in some way the stone to the club in order that a more effective blow might be struck. [126] The vegetable kingdom furnishes natural cords, flat stones with more or less cutting edges could be chosen and bound to the end of the club, and the earliest form of the battle-axe would be produced. With its formation the man-ape made another important step of progress and added greatly to his powers of offence. Stage by stage he was bringing his animal competitors under his control.

The formation of an axe or hatchet, however crude it may have been, would naturally lead to another step in advance. With it the ancestral man had passed beyond the possession of a weapon into the possession of a tool. The shaping of his clubs previously had been done by a rude tearing or hammering off of their twigs. These could now be cut off, and in addition the club might be wrought into a better shape. Manufacture had begun. Our ancestor stood at one end of a long line, at the other end of which we behold the steam-engine, the electric motor, and an interminable variety of other instruments.

Primitive manufacture was not confined to the shaping of wood. The shaping of stone followed in due time. If a tree branch could be made more suitable for its purpose by cutting it into shape with a rude stone axe or hatchet, a stone of better shape might be obtained by hammering. Doubtless the chipping effect of striking stone upon stone had been often observed before the idea arose that this could be made useful, and that where stones of the [127] desired shape were not to be found, the shape of those at hand might in this way be improved.

If we seek for some turning-point, some stage of progress, in which the man-ape fairly emerged into man, perhaps it would be well to select that which we have now reached, that in which the animal in question, which had hitherto used the objects of nature in their natural form, first gained the idea of manufacture and began to shape these objects by the use of tools. In truth, the dividing line between man-ape and man was imperceptibly fine. Various points of demarcation might be chosen, each founded on some important step in evolution. But among them all that in which the effort to convert the objects of nature into better weapons by the use of tools is perhaps the best, as it was probably the first step in that long process of manufacture to which man owes his wonderful advance.

With this early effort at manufacture, man had reached a stage in which he was first able to make a permanent record of his existence upon the earth—aside from that of the very infrequent preservation of his bones as fossil remains. A chipped stone is a permanent object. Even a very rudely shaped one bears some indications of its origin upon its surface, some marks pointing back to man in his early days. Unfortunately for anthropologists, natural agencies sometimes produce effects resembling those achieved by man's hands, [128] and some degree of skill in manufacture and well-marked design is necessary before one can be sure that a seeming stone weapon has not been shaped by nature instead of man. Within a recent period research for the evidence of early man in the shape of chipped stones has been diligently made, with an abundance of undoubted and a number of doubtful results. Some of these reach very far back in time, and if actually the work of man he must have lived upon the earth as a manufacturing animal for years that may be numbered by the million. Seemingly chipped stones have been found that belong to the remote Miocene geological age. With the latter are some scratches upon bones that also seem the work of tools. But these Miocene relics are questionable. They do not seem to surpass the shaping power of nature herself. Unless some more indubitable relics are found, we must place the advent of man as a tool-using animal at a much later date. How far back he may have existed as a man-like biped is another question, which we are not likely soon to solve.

It is scarcely necessary to pursue this branch of our subject farther. We have reached one end of a line of development, the succeeding course of which is well known. From the earliest rudely chipped stones and flints that are certainly the work of man, we can easily trace his progress upward through better examples of the chipped and later through those of the polished stone imple [129] ment, until the age of metal began. And with these stones have been found many other indications of the progressing powers of man, in the shaping of bone, the invention and use of a considerable variety of implements and ornaments, and the earliest efforts of art, as stated in a preceding section. There is no occasion to go into the detail of these steps of progress. When they are reached, this section of our work ends. We are concerned here simply with man's ancestor and man in his earliest stage of existence, not with man in his later course of development.

[130]

IX

THE FIRST STAGE OF HUMAN EVOLUTION

The question has often been asked, if man has descended from an ape ancestor why is it that no traces of this ancestral form have been found in a fossil state? If man has gone through such an extended course of development, why has he left no remains? This question, looked upon as unanswerable by many of those who ask it, is really of minor importance. A half-dozen answers, each of considerable weight, could easily be made to it. In the first place, it may be said that the absence of remains referred to is far from a single instance, but one out of thousands. It is generally admitted that the species of animals found fossil are very far from representing all the species that have existed upon the earth, and probably form but a minute percentage of them. In the second place, the remains of man's ancestor have not been sought for in its native locality, the tropical regions. In the third place, man belongs to the class of animals least likely to be preserved in the fossil state, since they dwell in the depths of forests and at a distance from the lakes and streams in whose muddy bottoms the remains of so many [131] animals have been fossilized. Another answer is, that of the various species of anthropoid apes that probably existed in the past, a few relics only of a single species have been found. If there were this one species alone, its number of individuals must have reached into the millions, yet of those hosts only a few fugitive bones are known to exist. There could not well be a more striking instance of the imperfection of the geological record. The sparse remains of Dryopithecus, the species in question, with some few other fossils of doubtfully anthropoid species, save us from a total blank, and open the vista to a myriad of active arboreal creatures which had their dwelling-place in the old-time European forests, but have almost utterly vanished from human knowledge.

These are not the only answers that can be made to the question propounded. Though the bones of the man-ape have not been found, relics of several stages of developing man exist. Most significant among these, until recently, was the celebrated Neanderthal skull, which in facial aspect departs widely from the ordinary human and approaches the simian type. More significant still is the Pithecanthropus cranium, indicative of an animal that stood midway between man and ape, a creature fully erect in posture, as its thigh bone proves, but with a brain that had attained but the halfway stage of development. In this notable find we seem to see man in the making, the [132] body already fully man-like, the brain advanced much beyond the stage of the ape intellect, but still far below that of man. It is the remnant of a creature significantly on the dividing line between man-ape and man.

So much for the response to the question as hitherto made. As the case stands, we are not obliged to stop at this point. Within the latter section of the nineteenth century discoveries have been made which fit in admirably with our argument. Rediscoveries, perhaps, we should call them, for they were imperfectly known in ancient times, but only recently have they fairly come within human ken. We refer to the Pygmy tribes of the African forests, not definitely offered hitherto as aids to the elucidation of this problem, yet which seem to adapt themselves closely to it, and certainly help essentially in filling the gap between civilized man and his ape-like ancestor.

We have already said that there appear to have been two separate and distinct stages in the evolution of man: one, that of his conflict with the animal world, ending in his mastery of the brute creation; the second that of his conflict with nature, ending in his mastery of the resources of the earth. Overlapping and succeeding the second there has been a third, that of the conflict of man with man, ending in the survival of the fittest of the human race. In the discussion of this problem, as hitherto made, these distinct stages of evolution, [133] with their intermediate resting stages, have not been recognized; argument being based on man as a whole, and no thought directed to the possibility that existing man may represent several separate processes of development, with broad lapses between. The argument we propose to offer is that man as he was at the completion of his first stage, that of the subjugation of the animal world, and before the beginning of the conflict with nature, still exists, the first derivation from the man-ape, living in the location and possessing much of the appearance and many of the habits of this ancestral form.

Late travellers in Africa have found more than trees and streams in the forest depths. They have found there a distinct and peculiar race of men, negro-like in many particulars, yet differing from the negroes in others, and specially marked by their dwarfish stature, which is indicated in the name of Pygmies, usually given them. These diminutive beings were known as long ago as the days of Homer, and their legendary combats with the cranes are spoken of by him in his poems. He was not aware of what is known now, that these forest dwarfs would disdain the cranes as antagonists, and are quite capable of overcoming the lordly elephant. In truth, they know no equals in the forest, and, while destitute of any knowledge of agriculture, are the most skilful, considering the primitive character of their weapons, of the hunters of the earth.

[134] The forest is the home of the Pygmy, as in all probability it was of the man-ape. He dwells in its deepest recesses, its moist and sultry depths, and pines when removed from his native realm in the heart of the tropic woods. In truth, he is almost as fully arboreal as was his tree-dwelling ancestor and as are his forest relatives, the anthropoid apes of to-day; not inhabiting the limbs of trees, indeed, but living under their shade, and forming the true man of the woodland, the nomad hunters of the vast equatorial forests. It must be said, however, that this is not wholly the case. There are tribes seemingly belonging to this race in South Africa who dwell in the open desert, but retain there, in great measure, the habits of their forest kin.

The first of modern travellers to see the Pygmies was Du Chaillu, in his journey through the African woodlands in 1867. He describes them as averaging four feet seven inches in height, their complexion of a pale yellow brown, the hair of their head short, but their bodies covered with a thick growth of hair, as if the loss of their ancestral covering had not been completed. The tribe seen by him was known as the Obongo, and dwelt in Ashango Land, occupying the forest region between the Gaboon and the Congo.

Dr. Schweinfurth, whose exploration extended from 1868 to 1870, was the next to meet these nomads of the forests, of whom he has given an [135] interesting description in his "Heart of Africa." He met with them in the country of the Manbuttoo, on the Welle River, between three degrees and four degrees north latitude. The tribe seen by him, known as the Akka, was made up of very diminutive individuals, none being over four feet ten inches high, and some only four feet. Their bodies were in due proportion to their height, so that they resembled half-grown boys in size.

The Akkas, as described by him, have large heads, huge ears, and very prognathous faces. Their arms are long and lank, the chest flat and narrow, widening below to support a huge hanging abdomen, the legs short and bandy, and the walk a waddling motion, there being a sort of lurch with each step. In this latter respect they recall the gibbon in its effort to walk. The gaping aspect of the mouth has a suggestive resemblance to that of the ape. They are also ape-like in their incessant play of countenance, twitching of eyebrows, rapid gestures of hands and feet, nodding and wagging of the head, and remarkable agility. Their skin is of a dull brown color, "like partly roasted coffee," and destitute of the covering of hair seen by Du Chaillu on the Obongos. The hair of the head and the beard is scanty and of woolly texture.

Stanley, who frequently met those forest dwarfs in his expedition for the relief of Emin Pacha, gives much information concerning them in his [136] "In Darkest Africa." He found, indeed, two types of dwarfs, one the Wambutti, who were of attractive aspect, having large, round eyes, full and prominent round faces with broad foreheads, jaws slightly prognathous, hands and feet small, figures well formed though diminutive, and complexion of a brick red hue. The other type, the Akka, he describes as having "small, cunning, monkey eyes, close and deeply set." One woman described by him had "protruding lips overhanging her chin, a prominent abdomen, narrow flat chest, sloping shoulders, long arms, feet strongly turned inward, and very short lower legs." She was "certainly deserving of being classed as an extremely low, degraded, almost a bestial type of a human being." The language of the Akka is of a very undeveloped type, and seems a link between articulate and inarticulate speech.

Stanley, in his journey down the Congo, heard many stories of the forest dwarfs, who were described to him as a yard high, with long beards and large heads. Other traditional accounts of them similarly speak of their long beards, though Stanley saw none answering to this description. The first individual seen by him in this journey was four feet six and a half inches high, and measured thirty inches round the chest. He was of a light chocolate color, with a thin fringe of whiskers, his legs bowed and with thin shanks, the calf being undeveloped. His body was covered with a [137] thick, fur-like hair, nearly half an inch long, in this respect agreeing with those described by Du Chaillu.

The Batwas, seen and measured by Dr. Ludwig Wolfe in the middle Congo basin in 1886, were of an average height of four feet three inches. They resemble the Akka in general appearance, and have longish heads, long narrow faces, and small reddish eyes. They bounded through the tall herbage "like grasshoppers" and were remarkably agile in climbing.

For several years past there have been rumors of a race of Pygmies in the interior of the Cameroons, but these reports were not verified until the year 1898, when the Bulu expedition of the German military force succeeded, with much difficulty, in seeing several individuals of this race, secured through the aid of a native chief. One woman was measured and proved to be just four feet high. The color was from chocolate-brown to copperish, except the palms, which were of a yellowish white. The hair was deep black, thick, and frizzled; the skull broad and high; the lips full and swollen. Like other Pygmy tribes, these are very shy, wandering from place to place in the forest, and avoiding frequented routes of travel. They are skilful hunters and collect much rubber, which they dispose of to the negro tribes.

In the same year Mr. Albert B. Lloyd made a journey in Central Africa, following Stanley's [138] route down the Congo. He was alone, with the exception of a few carriers, and had the good fortune of passing through the country of the Pygmies and that of the cannibals of the Aruwimi without conflict or injury, entering into cordial relations with both peoples. He journeyed for three weeks in the Pygmy forest and had excellent opportunities for examining its inhabitants.

After entering the great primeval forest Mr. Lloyd went west for five days without the sight of a Pygmy. Suddenly he became aware of their presence by mysterious movements among the trees, which he at first attributed to the monkeys. Finally he came to a clearing and stopped at an Arab village, where he met a great number of the diminutive nomads. "They told me," says Mr. Lloyd, "that, unknown to me, they had been watching me for five days, peering through the growth of the forest. They appeared very much frightened, and even when speaking covered their faces. I asked a chief to allow me to photograph the dwarfs, and he brought a dozen together. I was able to secure a snap-shot, but did not succeed in the time exposure, as the Pygmies would not stand still. Then I tried to measure them, and found not one over four feet in height. All were fully developed, the women somewhat slighter than the men. I was amazed at their sturdiness. The men have long beards, reaching halfway down the chest. They are very timid, and will [139] not look a stranger in the face, their bead-like eyes constantly shifting. They are, it struck me, fairly intelligent. I had a long talk with a chief, who conversed intelligently about their customs in the forest and the number of the tribesmen. Both men and women, except for a tiny strip of bark, were quite nude. The men were armed with poisoned arrows. The chief told me the tribes were nomadic, and never slept two nights in the same place. They just huddle together in hastily thrown-up huts. Memories of a white traveller,—Mr. Stanley, of course,—who crossed the forest years ago, still linger among them."

The discovery of these forest Pygmies has directed attention to the Bushmen of South Africa, a desert-dwelling race, long known though comparatively little regarded in their ethnological significance. They are now by many regarded as an outlying branch of the forest Pygmies, the chief difference being in the shape of the skull, which is rather long in the Bushmen, rather short in the Pygmies. These degraded wanderers inhabit an area extending from the inner ranges of the mountains of Cape Colony, through the central Kalahari desert, to near Lake Ngami, and thence northwestward to the Ovambo River. Into these, the most barren portions of the South African deserts, they have been driven by the encroachments of Kaffirs, Hottentots, and Europeans.

[140] They closely resemble the Akka tribes of the north, averaging about four and a half feet in height, and possessing deep-set,

crafty eyes, small and depressed nose, and a generally repulsive countenance. Their complexion is of a dirty yellow. Their hair grows in small, woolly tufts. In the vicinity of Lake Ngami, Livingstone found them to be of larger stature and darker color, while Baines measured some in this region who were five feet six inches in height. In disposition the Bushmen are strikingly wild, malicious, and intractable, while their cerebral development is classed by Humboldt as belonging to almost the lowest class of the human species.

Close in affinity with the Bushmen, and in various respects unlike the dark races around them, are the Hottentots, the original inhabitants of Cape Colony, a race of herdsmen who are much superior in culture to the degraded desert nomads. They are not dwarfish, being of medium stature, but they resemble the Bushmen in complexion, in which and in general cast of features they present some similarity to the Chinese. Their hair, like that of the Bushmen, grows in tufts, with spaces between, and they are like them in language, their method of speech consisting largely in a series of clicking sounds. Their manner of talking has been compared to the clucking of a hen, and by the Dutch to the "gobbling of a turkeycock." The Hottentots present every appearance [141] of being a developed branch of the Pygmy family, or the result of a cross between Bushmen and negroes.

These tribes of dwarfs, now extended throughout the equatorial forests and over the South African deserts, were probably once far more widespread, inhabiting much of the continent and reaching as far as Madagascar, where a branch of them, known as Kinios or Quinias, are thought still to exist. They extended north to the Mediterranean, and have left their representatives in Morocco in a tribe of dwarfs, about four feet high, who differ widely in appearance from all other people of that country. As to their origin, there is a diversity of opinion. Some anthropologists look upon them as a primeval race, distinct from the negroes, who came among them later. Professor Virchow, on the contrary, is of the opinion that their only important difference from the negroes is that of size, and regards them as the remains of a primitive population from whom the negroes have descended.

In a preceding section a statement was made as to what was the probable general appearance of the man-ape. It was based upon the physical aspect of the Pygmies, whom we hold to form the immediate derivative of man's ape ancestor, and to have made no radical change in personal appearance, if we may judge from the various ape-like characteristics which they still present. [142] Mentally they have made a very considerable advance, and have reached the stage of men of low intellectual powers; but while their brains have been growing their bodies have not greatly changed, and the marks of their origin are thick upon them. There has probably been little change in size, the diminutive stature and small bodily dimensions being in accord with their incessant activity, while the difficulties of traversing the thick growth of the tropical forest may have helped to keep them small. As it is, they are of about half the size of civilized man, the weight of a full grown adult male being probably not over ninety pounds.

Taking the Pygmies as a whole, it may be said that, though many of the Akkas are disproportionate in shape and tottering in gait, on the whole these people are well made, their protuberant paunch being probably a result of their habits of eating. Captain Guy Burrows says that a Pygmy will eat twice as much as would suffice a full-grown man, and that one of them will devour a whole stalk of bananas at a meal, with other food. Some tribes are described as physically and mentally degenerate, and prognathism is in many cases strongly declared, the lower part of the face having an ape-like contour, and the protruding chin, that feature peculiar to man, being very deficient. In their great abdominal development the adult Akkas resemble the children of Arabs and negroes. This, therefore, seems the retention of a primitive feature [143] which has become a passing characteristic in the more advanced types of mankind.

The Pygmies are not destitute of intelligence, and are capable of receiving some of the elements of education. Two of them were brought to Italy about 1875, who within two years' time learned to read and write and to speak Italian with much fluency. They showed themselves superior in school studies to European children of ten or twelve years of age, and one of them became somewhat proficient in music. In their habits they resembled children, being sensitive and impulsive, fond of play, and very quick in their motions. Their readiness in gaining the elements of education is in accord with experience in the case of other savages. It is when studies requiring abstruse thought are reached that the facility in acquisition of the savage races comes to an end.

With this consideration of the characteristics and habitat of the Pygmies we may proceed to a review of their habits. The weapons which they seem to have developed during their long upward progress, and to which their supremacy over the wild beasts of the forest is probably due, consist of two, the bow and arrow and the spear. The bow and arrow are small and insignificant in appearance, and would be of little value but for the poison which the Pygmies have somehow learned how to obtain, and which makes them dreaded, not only by beasts, but by men. Wherever found, from the deserts of [144] the south to the forest of the Welle and Aruwimi on the north, the poisoned arrow is a mark of affinity as decided in its way as their physical resemblance. Its wide distribution goes to indicate that it was the general weapon of the Pygmies ages ago, when, presumably, they had all Africa for their own, and ruled supreme over the animal world in that continent.

It is true, indeed, that the use of the poisoned arrow is not peculiar to them, but is a somewhat common possession of savage tribes in all parts of the earth. This makes it quite possible that it was not original with the Pygmies, but was derived by them from other tribes. On the other hand, in view of its great value in giving them supremacy over the lower animals, it may well have been a primeval Pygmy invention, and these tribes the original source of its existing wide distribution.

They possess more than one poison; one being a dark substance of the color and consistence of pitch, which is supposed to be made out of a species of arum. It is laid in the splints of their wooden arrows, or spread thickly upon their iron arrowheads, when they possess these. Another poison is of a pale glue color, which is supposed by Stanley to be made of crushed red ants. When fresh these poisons are deadly, producing excessive faintness, palpitation of the heart, nausea, and deep pallor, soon followed by death. In Stanley's experience one man died within a minute, from a mere [145] pin prick in the breast. Others lived during different intervals, extending up to one hundred hours. The difference in virulence seems to have depended on the degree of freshness of the venom, which apparently lost its strength as it became dry.

The possession of a weapon so deadly as this, together with the agility and daring and the unerring marksmanship of the forest dwarfs, seem sufficient to give them absolute control of the animals of the African wilds. The lion, the elephant, and the buffalo, the largest and fiercest of the beasts of field and forest, are powerless before the virulent venom of the arrows of the Pygmies, and doubtless for ages they have held dominion as the fearless rulers of wood and wild. Captain Burrows says of the skill with the bow of the Pygmy that "he will shoot three or four arrows, one after the other, with such rapidity that the last will have left the bow before the first has reached its goal."

The bow and spear are not their only means of obtaining food. They have certain of the arts of the trapper, perhaps original with them, perhaps borrowed from their larger neighbors. They sink pits in the pathways of their game, covering them with light sticks and leaves and sprinkling earth over the whole. They build hut-like structures, and lay nuts or plantains beneath, for the purpose of tempting chimpanzees, baboons, or other apes. A slight movement causes the hut to fall on the incautious animals. Bow traps are placed along [146] the tracks of civets, ichneumons, and rodents, which snap and strangle them. The Pygmies do not hesitate to attack the elephant, spearing it from beneath, and hunting it for its ivory, which they trade with the settled tribes. In short, they are of unsurpassed agility, and are the best of woodsmen and hunters, their skill being taken advantage of by the settled tribes, who trade with them vegetables, tobacco, spears, knives, and arrows for meat, honey, the feathers of birds, the ivory of the elephant, and other forest spoil. So destructive are they of game that they would soon denude the surrounding forest if they stayed long in one spot, so that they are compelled to move frequently. Schweinfurth speaks of them as cruel and fond of tormenting animals.

They serve the settled natives in other ways, acting as scouts and informing them of the coming of strangers while still distant. Every forest road runs through their camps, their villages command every crossway, and no movement can take place in the forest without their knowledge, while they are adept in the art of concealment.

The superior woodcraft, the malicious disposition, and the poisoned arrows and good marksmanship of these forest folks make them formidable enemies, and the settled tribes hold them in dread and are glad to keep on good terms with them. Yet they find them much of a nuisance, since their dwarfish neighbors claim free access to their gar [147] dens and plantain fields, where they help themselves to fruit in return for small supplies of meat and furs. In short, they are human parasites on the larger natives, who suffer from their extortions, yet fear to provoke their enmity. Burrows says that they will never steal, but that they pay very inadequately for the plantains they take, leaving a very small package of meat in return for an ample supply of food.

The Pygmies build their camps two or three miles away from the negro villages, living in groups of sixty to eighty families. A large clearing may have eight to twelve of these Pygmy camps around it, with perhaps two thousand inmates. Their dwellings are of the shape of an oval cut lengthwise, and are built in a rude circle, the residence of the chief occupying the centre. The doors are two or three feet high. On every track leading to the camp, at about one hundred yards' distance, is a sentry house large enough to hold two of the little folks, its doorway looking up the track from the camp. While wandering in the forest they build the flimsiest of leaf shelters.

The intelligence of the Pygmies is of a very low order. In the arts which they have been developing for ages they are experts, they are thoroughly familiar with the habits of animals, and as hunters they are unsurpassed. But in intellect they are decidedly lacking. They are destitute of agriculture, possess no animals except a few dogs, and [148] have none of the elements of culture. The Bushmen, for instance, can count only up to two; all beyond that is "many." Yet this low tribe of desert nomads is, as we have said, skilled in the art of drawing, its sketches of men and animals being widely distributed through Cape Colony.

The Pygmies seem greatly lacking in the social sentiments. Burrows, in his "Land of the Pygmies," says that they do not possess even the most ordinary ties of family affection. Such common and natural feelings of affinity as those between mother and son, brother and sister, etc., seemed to be wanting in them.

It is a fact of great interest that the Pygmy race does not seem confined to Africa, for tribes of men resembling the Pygmies in stature and in various other particulars are found in widely removed localities, as in Malacca, the Andaman Islands, and the Philippine Archipelago, while there are indications that they once spread widely over this island region of the earth. Those of the Philippines, known as Negritos or Aetas, have been somewhat closely observed and may be briefly described.

The Negritos are similar in stature to the Pygmies of Africa, the men averaging four feet eight inches high, and they are like them in general appearance. They are darker in complexion, some being as sable as negroes, and all of them darker than the African Pygmies. Their features are coarse and [149] ill-shaped, their nose depressed, lips full, hair black and frizzled. In body, like the Pygmies, they are thin and spindle-legged. The calf of the leg is not developed in any of these dwarfish people. The Negritos possess one marked and significant characteristic,—the separation of the great toe. This, while it has not the full power of movement shown in the apes, is much more separated from the others than in the whites, and can be readily used in grasping. By its aid the Negrito can not only pick up small objects, but can descend the rigging of a ship head downward, holding on like a monkey by his toes. It may be said that among uncivilized and barefoot people the great toe is usually very mobile. The artisans of Bengal can weave, the Chinese boatmen can row, with its aid, and it adds much to facility in climbing.

The Negritos wear little clothing, have no fixed abodes, and pass a wandering life in the forests, living on game, honey, wild fruits, roots of the arum, and other forest food. Their weapons consist of a bamboo lance, a bow of palm wood, and a quiver of poisoned arrows. It is certainly a striking fact that, wherever found, from South Africa to the Far East, the Pygmy tribes possess the art of poisoning their weapons. This art is not practised by the surrounding peoples, and is the strongest evidence of a community of origin. It seems to point back to a remote period when the Pygmy peoples spread far through the tropics [150] of the Eastern hemisphere, though in the region now under consideration they have almost vanished through the assaults of the Malays.

The Negritos are very alert physically, being remarkably fleet of foot, while they can climb like monkeys. They live in groups of about fifty families, shelter being obtained by a simple erection of sloping poles and leaves, though in their more settled locations they built bamboo huts like those of the Malays. They are a short-lived race, seldom living more than forty years. Mentally, they are stupid and apparently incapable of improvement, seeming to stand at the foot of the human scale. Attempts to instruct them have been made, but all proved failures. Efforts to make agriculturists of them have proved similarly futile. They are hereditarily hunters, and hunters they are likely to remain.

The only Eastern locality of which the Pygmy race remained in full possession until recent times is that of the Andaman Islands. This is no longer the case. Great Britain made a penal settlement of these islands after the mutiny in India, and as a consequence the Mincopies, as their native inhabitants are called, have begun to disappear. These islanders are rather taller than the Philippine Negritos, ranging from four and a half to five feet in height, but otherwise there is a somewhat close resemblance between them. Their color is dark brown or black, their hair woolly, and inclined to [151] grow in tufts, like that of the Bushmen. The head, though large in proportion to the body, is really very small and of low cranial capacity. That of the men is only 1244 cubic centimetres, as contrasted with 1554 cubic centimetres of a large number of male Parisians measured by Broca. That of the women differs in the same proportion. Flower says that the Mincopies rank lowest among the human races in this respect; but it must be remembered that the brain usually decreases in size with decrease in stature.

Small as these islanders are, however, their strength is relatively great. They use with ease bows which the strongest English sailors cannot string, though practice may have much to do with this facility. And they can send arrows with a force that seems out of accord with their size. Their agility is remarkable. Travellers speak of the speed of the bullet in describing their running—doubtless with some exaggeration. Their senses are strikingly acute. It is said that they can distinguish fruits by their odor when hidden in the foliage of the jungle, and have wonderful powers of sight and hearing. As in the case of the Aetas, their life is short, though the age of puberty is nearly as great as with us. Fifty is extreme old age with these people, and twenty-two is said to be their average length of life.

Mentally, they are at a low level, the lowest, in the opinion of Owen, among the races of mankind. [152] In counting they have words for only one and two, but can count up to ten by touching the nose with each of the fingers in succession, saying each time, "this one also." Their language is of a primitive type, and in various respects they manifest low intelligence. Yet, as in the case of the Akkas mentioned, they can be taught to the level of other children of twelve or fourteen years. Their mind, in the opinion of Dr. Brander, seems rather to be asleep than incapable. One child was taught to read and write, and to speak English fluently, and gained some knowledge of arithmetic; and this was not an exceptional case.

It does not seem at all remarkable, when we consider the ease with which monkeys can be taught many arts and acts new to them, that those dwarfish men, like other savages, greatly superior as they are in brain power to the apes, should be capable of acquiring the minor elements of education. It is not what they can be taught, but what they have taught themselves, that we must consider in assigning them to their comparative place in intellectual development. In this respect the Mincopies are on a very low plane. They have not even acquired the art of making a fire, though this is almost universal with mankind. All they know is how to keep a fire alive, and in this they are very assiduous. It is probable that they may have obtained fire at first from volcanoes on neighboring islands.

[153] They are lacking, like the Pygmy races in general, in the art of chipping stone, one of the earliest arts acquired by man. Their only means of shaping stone is to put it into the fire until it breaks or splinters, when they can use the sharp splinters for their purposes. They are quite destitute of the art of drawing, and have no means of communicating their thoughts except by speech.

Yet with these deficiencies, they have made some progress in the industrial arts. They make wooden vessels, and can produce pottery which stands the fire and in which they cook most of their food. They make nets of considerable size, which they use to fish with in the narrow streams. They have arrows and harpoons, whose points are fastened to the shaft by a long cord. The fish or land animal struck unwinds this cord in trying to get away, and its speed being checked by the shaft which it drags along, it is easily caught.

The Mincopies possess boats, and these seem to have been early possessions of the Negrito populations, by whose aid they were able to migrate from island to island. Their canoes have nautical qualities which have astonished English sailors. At one time they were probably bold and daring fishermen and navigators, until driven to the forests and mountains by the invasion of the Malays.

As the Pygmies were in all probability the aborigines of Africa, so the Negritos appear to have been the aboriginal people of the Eastern islands, if not [154] of India. Quatrefages, in his work "The Pygmies," finds reason to believe that even at the present day traces of them, pure or mixed, can be found from southeast New Guinea to the Andaman Islands, and from the Sunda Islands to Japan. On the continent their range extends, according to him, "from Annam and the peninsula of Malacca to the western Ghauts, and from Cape Comorin to the Himalayas."

In one part of India the Negrito-like population are called *Banderlokh* (literally "man-ape") by the neighboring tribes. The Semangs of Malacca are jet-black in color, with thick lips, flat nose, and protruding abdomen. In regard to the characteristic of prognathism, it is possessed in various degrees, the most pronounced instance being seen in the photograph of one of the Kalangs of Java, a tribe which has recently become extinct. The face of this individual is strikingly ape-like in profile.

Everywhere that these dwarfish people are found, whether in Africa, India, or Malaysia, they present the appearance of being an aboriginal race, now largely annihilated by the incursions of larger and better-armed people, but once widespread and numerous. As to their place of origin, whether in Africa, India, or the island region, it is useless to speculate, as the facts on which an opinion could be based are not known. Wherever found they are in close relation to the black races, the negroes of Africa, the Papuans of Polynesia,

[155] and evidences of a considerable degree of mixture of races exist. This is especially the case in Polynesia and India, where the Negritos appear to shade off into the full-sized blacks through an intermediate series of half-breeds.

Yet one fact of ethnological importance needs to be mentioned. The Negritos and Pygmies are everywhere brachycephalic, or short-headed, with the exception of the Bushmen, who are dolichocephalic, or partially so. Negroes and Papuans are strongly dolichocephalic. In this respect the Pygmy peoples agree more closely with the short-headed Mongolian or yellow races than with the long-headed negro or black races, though in general features they come near the latter.

In truth, this race of dwarfs may be the primitive stock from which the Mongolians branched off on the one hand, and the Negroes on the other, since they are in some measure intermediate between the two. Latham says of the Rajmalis mountaineers, "Some say their physiognomy is Mongolian, others that it is African." Quatrefages is strongly of the opinion that the negro is of Indian origin, and reached Africa through migration. He bases his opinion on the negroid characters of existing tribes in India, Persia, and elsewhere in Asia, and on the similar characters of the aboriginal Polynesians. As regards the Pygmies, they probably spread over the whole of this section of the earth at a period of remote antiquity, and very long ago [156] developed the racial differences which appear to exist between separate tribes. Distinctions of this kind can be seen in the East, and a marked one is pointed out by Stanley between the Wambutti and the Akka, as already stated.

Wherever found the Pygmies are hunters, usually making the deep forest their home, and are masters through their agility, cunning, and deadly weapons of the whole world of lower animals. Physically they are probably not far removed from the man-ape, their remote ancestor, for they retain various ape-like characters, as in aspect of face, shape of body, occasional hairiness, diminutive size, shortness of legs, imperfect development of the calf, occasional waddling gait in walking, and the other particulars above pointed out. There are certainly abundant reasons for believing them to be, as we have suggested, the final result of the first great conflict in the evolution of man, that with the lower animals.

This assured mastery once gained, the occasion for further development of this people ceased while they remained in the forest habitat which they had inherited from their ape ancestors. Here the problem of food getting was fully solved and there was nothing to instigate any new step in evolution. The period of conflict ended, a period of rest supervened, and, so far as the Pygmies are concerned, this period still continues. Though later races, their probable descendants, have left [157] the forest and set up new stages of development through new conflicts with adverse conditions, the Pygmies remain in their resting state, and, if left to themselves, might continue in this state for ages in the future as they have done for ages in the past. As the case now stands, however, annihilation threatens some of them, while educative and other influences from without may bring to an end the physical and mental isolation of the others.

In considering the Pygmies as they exist to-day, in fact, it is impossible to say how far their habits and possessions are original with themselves and how far they have been derived from others. There can be no question that they have been influenced by the customs of surrounding peoples of higher culture, and that they have received implements and methods from without. To get down to the pure Pygmy, as an outcome of evolution within himself, we would need to strip off all these adventitious aids, if we could distinguish them from the conditions native to the race, and thus behold him as he was before he fell under the influence of men of higher grade. Were it possible to isolate him in this way, and present his original self, we should have before us an ethnological specimen of the highest interest and importance, as the ultimate resultant of the first great stage in the evolution of man from his ape ancestor.

[158]

X

THE CONFLICT WITH NATURE

It has been a frequently debated question whether man comprises a single species or two or more species of animal descent. If a line be drawn from the Gold Coast in tropical Africa to the steppes of Tartary in central Asia, it will present two markedly distinct races of men at its two extremities. At its southwestern end we find the most long-headed, prognathous, frizzly-haired, dark-skinned race of mankind. At its northeastern end is the most round-headed, orthognathous, straight-haired, and yellow-skinned race. Midway between these appear intermediate peoples, with heads round, oval, or oblong, hair straight or curly, skin fair or dark, faces upright or protruding, men possibly, to judge from their physical character, a result of the amalgamation of these two distinct races.

These differences may be the result of original difference in species or may be due to climatic and other influences of nature. Some writers accept the one view, some the other, and neither is sustained by any great weight of facts. The Pygmy race presents somewhat similar differences. Usu [159] ally round-headed, these small men are in some instances long-headed, while such marked distinctions appear at times that Stanley classed two neighboring tribes as separate races. Here they present features of the Mongolian, there they are similar to the Negro. This goes to indicate that the distinction between the Negro and the Mongolian began far back in time, but it does not prove that it is the result of original difference in species, or that two distinct forms of ape separately developed into man. While this is quite possible, the theory of a single species has been most widely accepted. The chief writers on the subject think that the differences arose during that undeveloped stage of mankind when resistance to the transforming influences of nature was still weak, and when the structure of the human frame may have yielded readily to agencies which would have little or no effect upon it now.

Of one thing we can be sure, which is that there was a wide migration of the apes in remote times. Leaving the tropics, many species spread to the north, extending into Europe, which at that time seems to have been connected by land bridges with Africa, and spreading far through Asia. There was probably nothing at that time in atmospheric conditions to check such a migration. The Tertiary climate of Europe is believed to have been quite mild. And the ape family is by no means necessarily confined to warm regions. Monkeys [160] are found to-day at high elevations on the mountains of India, enduring the chill of ten thousand feet of altitude.

Of the migration to Europe abundant evidence exists, fossil remains of monkeys having been found in many localities of that continent. Among these residents of early Europe was at least one representative of the anthropoid apes, the fossil species known as Dryopithecus, from the middle Miocene deposits of St. Gaudens, France. This species, apparently most nearly allied to the chimpanzee, was taller than any existing ape. Two or three other fossil remains, possibly of anthropoid apes of smaller size, have been found, and Europe seems to have been well supplied with apes of a considerable degree of development at a remote geological period. Among those may have been the form we have designated the man-ape, the ancestor of the human race, though no fossil relic attributable to such a species has been recognized.

Coming down to a much lower period, we begin to find traces of man, first in his rudely chipped and later in his polished stone weapons and tools. And the bones of man himself appear, extending through what is known as the Quaternary or Pleistocene period. Nearly all these remains have been preserved by the art of burial, a fact indicating some degree of mental progress, though their residence in caves and the rudeness of their implements are evidence that the race was still low in culture.

[161] An interesting fact in connection with these ancient human remains is that most of them indicate a small race, with narrow skulls and prognathous jaws, recalling the Pygmies in general structure. This rude and small race continued until a late period of prehistoric time. It extended down from the cave bear and mammoth period through the later reindeer period, as is proved by discoveries made in the caves of the Belgian province of Namur. And there is good reason to believe that it continued into the age of bronze, for the small size of the handles of bronze weapons show they must have been intended for men with small hands.

These diminutive people seem to have been not over four feet eight inches high. They were not alone, however. Men of normal height were in Europe with them. The northward migration of the Pygmies seems to have been accompanied or followed by that of a full grown people. Yet the Pygmies have held their own in Europe as in Africa, with certain modifications. In Sicily and Sardinia, which form part of a supposed former land bridge between Africa and Europe, a small people about five feet high still exist, whom Dr. Kollman looks upon as representing a distinct race, the predecessors of the tall Europeans. In the Lapps of northern Europe we possess another small race, possibly the lineal descendents of the Quaternary Pygmies. Everywhere the small man has been [162] forced to retire into forests, deserts, and icy barrens before the taller and stronger man. The folk-lore of Europe is full of traditions of a race of dwarfs, and its conflict with men of larger mould, and there are various indications that this race was once widespread.

What has been said here of the migration of man into Europe and his development in that country is preliminary to a consideration of the second great stage of human development, that due to the conflict with nature. The conflict with the animal world appears to have ended in the production of a dwarfish, forest-dwelling variety of man, in the lowest human stage of mental evolution. The conflict with nature ended in the development of a full-sized variety of man, dwelling largely in the open country and much superior in intellect, as indicated by his higher powers of thought and advanced degree of organization.

The conflict with nature took several forms, in accordance with the conditions of the several regions inhabited by man. Its result was to subdue nature to the use and benefit of mankind, and the methods, in the tropical localities of original man, consisted in the reduction of animals to the domestic state and a similar domestication of food plants. In other words, one of its early stages was the development of the herding habit, while a far more important one was that of the appearance of the agricultural industries. In Europe a third [163] and still more vigorous influence supervened, that of the conflict with cold and man's gradual adaptation to the conditions of a frigid climate.

If the nomad dwarfs were the aboriginal men, all later races must have developed from them. While remaining in the forest and retaining their primitive habits, the Pygmies presented an instance of arrested evolution. For a new development to begin it was necessary to abandon the old locality and with it the old habits, and this they probably began to do at a remote period. When, indeed, the earth was their dominion, there was no reason for their remaining restricted to a forest residence, as they have been since the larger races took possession of the open country. We do not need to go back far in time in the East to find the Pygmy race in full control of the Philippine and other islands, and probably of Malacca and parts of Hindostan. Their present restriction and partial extermination have been due to the incursions of the warlike Malays. The Andaman Mincopies remained undisturbed until a recent date, and added fishing to their hunting pursuits. And the canoes which these islanders now possess were probably the invention of their race, and furnished the means by which the aborigines spread from island to island of those thickly studded seas.

In Africa the only existing indication of a migration of the forest folk into the open country is found in the Bushmen and Hottentots of the far [164] south. The former, confined to the desert, remain nomad hunters and present no step of advance beyond the Akka and other equatorial tribes. The Hottentots, on the contrary, have made an important step of progress. While still nomads and addicted to hunting, they have domesticated cattle and sheep and become essentially a herding people, though mentally the lowest race of herders on the face of the earth.

With this change in habits, the Hottentots have significantly increased in stature. While still of medium height, they are considerably larger than their Bushmen kindred, to whom they present a close resemblance in other respects. This increase in size is a common result of a change in habits which insures a fuller supply of food with less strain upon the muscular organization in obtaining it; a fact of which the lower animal world is full of illustrations. The life of the forest and desert hunters is one of incessant activity, and their food supply is precarious. The Hottentots, on the contrary, take life easily and are inclined to indolence, their herds supplying them with food in abundance with little exertion. They retain enough of the primeval strain to be fond of hunting, and while thus engaged display the activity of their ancestral race, but ordinarily they pursue an idle, wandering life, and their increase in size may well be a result of their change in habits.

The Hottentots, while still low in the human [165] scale, are mentally a stage in advance of the Bushmen, they having a more developed social organization and superior powers of thought. The latter is indicated by their myths and legends, of which they have a considerable store, though they are in great measure destitute of religious conceptions, such religion as they possess taking in great part the primitive form of ancestor worship. Under the influence of Europeans they are gradually abandoning their old habits and adopting those of civilized life, but while improving in social and industrial conditions there is little evidence of intellectual advance.

The development in method of food-getting displayed by the Hottentots was really but the completion of the old battle for dominion with the animal host. It consisted in subjecting some of the docile herbivora more fully to human mastership. The hunter has to do with hostile beasts, victims but not servants of man. The herder has reduced some of these animals to servitude, and no longer has to overcome them through the arduous labors of the chase. He is able to obtain, as we have said, more food with less exertion, a larger population can live in a limited district, and the beneficial effects upon the mind of a closer social intercourse are shown.

But the most important event in this stage of evolution was the subjection of the plant world to man. For ages of interminable length this was [166] not thought of. Fruits and other vegetable products formed part of man's food; but these were the growth of wild nature, and the plant world was left to its own will, with no effort to bring it under human control. There is nothing to show that the idea of agriculture ever entered the mind of a Pygmy. Of the plants surrounding him, far the greater number were useless for food, only the few were available; but the conception of favoring the few at the expense of the many apparently never occurred to him. There is, indeed, some crude and simple agriculture pursued by a few of the Negritos of Luzon, but evidently as an imitation of the Malay agriculture or as a result of direct teaching, certainly not as an original conception. The conflict of the Pygmies with nature has been confined to the animal world, and reached its highest level in the herding industries of the Hottentots.

Where and when the subjugation of the plant world began it is impossible to say. It very probably had its origin in the fertile open lands of the tropics. But that it originated in the central region of Africa, or that the agriculturists of that region were of native origin, are both subjects open to question. The forest folk may have spread into the open country, there developed a crude agriculture, favored the growth of food plants at the expense of useless shrubs and trees, and gradually advanced in this new form of industry. This would be in accordance with the opinion of Vir [167] chow, who looks upon the negro as the descendant of the Pygmy. No great change was necessary to convert the one into the other. The Pygmy is negro-like in cast of countenance and bodily formation. He differs in size, in complexion, and in shape of head. But new conditions may have given rise to these differences. The fierce suns of the African lowlands may well have caused an increased deposit of pigment, changing the yellowish hue of the Pygmy to the deep black of the negro. An increase in size is a natural result when exertion diminishes and food increases. And a tendency for the head to change from the short to the long shape is shown in the Bushmen.

On the other hand, certain anthropologists, of whom we may name Quatrefages, take an opposite view, and believe that the negroes migrated from Asia or the Eastern islands to Africa, being, like the negro-like Papuans, descendants of the sable or dark brown Negritos of the East. In this case agriculture may have originated in Asia and have been brought by migrants to Africa. All we know historically concerning it is that the earliest traceable seats of agriculture appear to have been the fertile valleys of India, Babylonia, and Egypt. But the known culture of the earth in these regions goes back only a few thousands of years, while for the first crude stages of agriculture we must probably measure years by tens of thousands.

The degree of subjection of nature to man's [168] needs, as displayed in tropical agriculture, was comparatively small, and its effect on the development of the human intellect, while important, was limited. It had the highly useful result of a great increase in population, the growth of village and town life, an advance in social relations, and the beginning of political relations. New implements were needed, better houses were erected, the settled condition of the people gave rise to direct efforts at education, and added the important element of commerce, in its earliest form, to the industries of mankind. The result must have been a fresh start in the development of the intellect, though one that probably soon reached its culminating point in the central tropics.

The highest results of the development of agriculture in tropical countries, unaided by secondary influences, seem to have been those existing in the highly fertile regions of Egypt and Babylonia at the opening of the historical period. The density of population in those countries, due to their prolific production of food stuffs, gave rise to considerably developed political and social institutions, and laid the foundations for a great subsequent advance under the influence of warfare, invasion, and the other more potent causes of human progress. Only for such ulterior influences the agriculturists of these countries would perhaps to-day remain dormant in the stage of mental progress they had attained ten thousand years ago.

[169] In considering the existing conditions of the forest nomads and the African agriculturists, it is not safe to credit them with the origination of all the arts and implements they possess. The negroes, for instance, have been for ages in more or less close association with the Pygmies, and may have taught them many things which they would not have attained through their own limited powers of thought. The bow and poisoned arrow are very likely original with them. They possess this weapon throughout the wide range from the African Hottentots to the Philippine Negritos, while it is not a weapon of the surrounding peoples. The spear is probably also original. The same cannot safely be said of their traps and snares for game. These seem beyond their power of invention, and may well have been taught them by the negro tribes. Their habitations, aside from the mere leaf shelters, had probably a similar origin. In Africa the huts doubtless had their model in those of the negroes. In the Philippines they are pile-supported bamboo huts of the pattern of those of the Malays. If, then, we take from the forest folk the arts taught them or imitated by them, we reduce them to a very low

level of intellect and a remarkable paucity of products from their own powers of thought.

Similar reasoning may be applied to the settled natives of Africa. For thousands of years past they have been in contact on their northern borders with civilized peoples, numerous immigrants [170] have made their way into the country, and a considerable degree of amalgamation has very likely taken place. We cannot, therefore, safely credit them with all the arts and implements they possess nor with all their political and social progress. No doubt much came to them from without, much was taught them from within, and a mixture of blood with superior races may have aided considerably in improving their stock. We are justified, then, in their case as in that of the Pygmies, in believing that their stage of mental and social development is only in part original with them, and is largely due to the influences of education and amalgamation.

The pure negro is not a very numerous element of the population of Africa. He stands in a measure intermediate between the nomad Pygmies of the forest and the desert, and the mixed races who may be called negroid but cannot strictly be called negro. With their foreign blood, most of these have obtained foreign arts and elements of culture, and stand at a distinctly higher physical and mental level than the unamalgamated negro.

For the pure or nearly pure negro we must seek the lowlands of the Guinea coast, the seat of the most pronounced existing negro type. Other localities are in the region of the Gaboon, along the lower Zambesi, and in the Benue and Shari basins. Here we find the true native African, a race strikingly uniform in aspect, and, next to [171] the Pygmies, the lowest in physical characteristics of mankind. The features of structure in which the negro appears to occupy a position intermediate between the white man and the man-ape—lower than the former and approaching the latter—are the following: First, his abnormal length of arm, which averages about two inches longer than that of the Caucasian, and, when in the erect position, sometimes reaches the knee-pan, being little shorter proportionately than that of the chimpanzee. Second, his prognathism, or projection of the jaws—his index of facial angle being about 70, as compared with the Caucasian 82. Third, his weight of brain—

average European 45 ounces, negro 35, highest gorilla 20. Fourth, his short, flat, snub nose, deeply depressed at the base, wide and with dilated nostrils at the extremity. Fifth, his thick protruding lips. Sixth, his high and prominent cheek bones. Seventh, his great thickness of cranium, which resists blows that would break the skull of an average European. Eighth, the weakness of his lower limbs, the broad, flat foot and low instep, the projecting heel and somewhat prehensile great toe.

These characteristics the negroes possess in common with the Pygmies and the Negritos. Others of less significance could be named. One important character is that of the cranial sutures, which close much earlier in the negro than in higher races, thus checking the development of the brain while [172] the body is still growing. To this many ascribe the mental inferiority of the negro race. A close observer records, as a result of long observation on the plantations of the southern United States, that "the negro children were sharp, intelligent, and full of vivacity, but on approaching the adult period a gradual change set in. The intellect seemed to become clouded, animation giving place to a sort of lethargy, briskness yielding to indolence." This is very probably the case with the Pygmies, who similarly reach a mental limit beyond which they cannot advance; but this limit is set in the adult period. In other words, the adult Pygmy is on the mental level of the negro child. If the African Pygmy is as short lived as his Eastern congener, he does not survive, as a rule, many years beyond the age of adolescence, and continues in a stage of childhood, mentally considered, until death.

The conclusion to be derived from this interesting fact would appear to be that the negro has made a distinct and important advance mentally beyond the Pygmy, reaching at adolescence the limit of mental evolution which the Pygmy reaches at death. But the negro stops here, or goes little beyond this limit. His cranial sutures close, the growth of the brain is arrested, and the development of his mind comes to an end. In the white the brain continues to expand, and the closing of the sutures takes place later in life. Probably the [173] latter is a result of the former, mental development having overcome the tendency of the sutures to close in early life. It may be further said of the negro that, mentally, he is emotional far more than intellectual, and unmoral rather than immoral, he being apparently incapable of comprehending the moral conceptions of advanced man.

If we seek the Malaysian and Australasian region of the Eastern seas, we find there another branch of the negro race, similarly in contact with, and apparently derived from, a Pygmy stock. This Papuan race of blacks covers a wide island region, but, like the African race, has become greatly modified by mixture with alien peoples, largely of Malay origin. Its purest type is to be found in New Guinea, where it approaches the negro in general character, though with distinctive features of its own.

The Papuan is of medium height; fleshy rather than muscular; color a sooty brown; forehead high, but narrow and retreating; nose sometimes flat and wide at nostrils, but oftener hooked with depressed point; lips thick and projecting; high cheek bones; prognathism general; hair black and frizzly. He is negroid in appearance, and is said to resemble the African of the coast region opposite Aden.

We need not pursue this subject further. It will suffice to offer the general conclusion that the negroid race, while, through its change of habits [174] from the hunting to the agricultural status, it has made an advance both mentally and physically beyond the Pygmy aborigines, does not appear to have advanced greatly in either particular, the negro reaching a mental limit at a low level, and being arrested physically while still possessing marked characteristics of the man-ape.

For the higher development of man, under the stress of a more energetic conflict with the conditions of nature, we must seek the continent of Europe, whose human inhabitants had not only to subdue the wild beasts and teach the earth to bring forth wholesome food in place of useless plants, but also to battle with wintry climates, and overcome the adverse influences of cold, sterility of soil, and other hostile conditions of the northern zones.

One of the chief problems of biology has long been that of the production of new varieties and species of animals as an effect of gradual variation in structure. This is believed to be ordinarily due to changes in the conditions of nature, animals and plants which have made accordant changes in structure being preserved, those which have not changed in accordance with the new conditions perishing. Where the conditions of nature remain uniform, species may persist for long ages unchanged, though even in the latter case changes in structure are apt to occur, since variation in species is not wholly dependent upon external changes. [175] To a considerable extent it is due to causes existing within the organism itself, fortuitous variations being occasionally preserved when not out of harmony with the state of affairs prevailing in the external world. Or variation may occur through the establishment of new relations between the species inhabiting some locality while inanimate nature remains uniform, or through migration into new inanimate or animate surroundings. Variations, in short, may arise under the influence of any change in the general environment which renders necessary adaptive changes in structure. But this adaptation in some cases takes place in the mind, new actions or methods of meeting the contingency being adopted which render physical changes unnecessary. The problem is a highly complicated one, and no doubt many causes have to do with the multiplicity of effects.

There have very likely been many occasions where the changes in structure took place rapidly, in consequence of sudden variations in natural conditions. Such rapid changes in conditions necessarily exert a severe stress or strain on organisms, either destroying them or causing an equally rapid adaptation, physical or mental. In such instances it is likely that many species perish, the change demanded being too great; others escape by migration to better fitted localities; and others, more mobile or less affected by the change, survive through adaptive variations.

[176] Of such periods of strain upon organic nature we know of only one in recent geological times, that known as the Glacial Age, the vast variation in climate which took place when the ice of the Far North flowed down in mighty billows over northern Europe and America, burying everything beneath its crushing weight, and bringing many forms of life to a sudden and untimely end. No doubt a considerable number of species of animals and plants perished before this frightful invasion. A notable instance among these was perhaps that of the American horse, which disappeared at about this period. Other species survived by a retreat to more tropical regions, to return after the invasion had spent its force. Still oth-

ers may have survived by adapting themselves to the changed conditions, emerging as new species or well-marked varieties.

Among the beings which passed unscathed through this extraordinary change in climate was apparently man. And it seems safe to affirm that man's contest with the glacial conditions, whose force was exerted upon his mind instead of on his body, was one of the most potent influences in the evolution of the human race. Man entered the contest at a low level of mental development; he emerged from it at a comparatively high level.

No one to-day questions that man was an inhabitant of Europe during the Glacial Age. The [177] proofs of this are too numerous and positive to be doubted. He may have inhabited America in the same period, though of this there still remains some doubt. Claims have been made of the discovery of evidences of man in Europe long before the glacial epoch, reaching as far back as the Pliocene and even the Miocene Age. But these claims have not been established beyond question, and the earliest generally acknowledged traces of man are confined to glacial Europe.

Yet we are forced to acknowledge that if man existed in Europe during the prevalence of the ice age, he, or his ancestor, must have been there before that period. It is absolutely certain that no animal accustomed to tropical conditions would have chosen this period of extreme cold to migrate from the warm tropics to the frozen north. The fact that man was in Europe during glacial times is the very strongest evidence that he reached there during the milder preceding period, when a genial and uniform climate is believed to have prevailed throughout southern and central Europe. If we could accept as fact the seeming very ancient evidences of man's handiwork, we would be obliged to consider him an inmate of Europe long ages before the glacial epoch.

If, as there is reason to believe, the man of Africa at that remote period was the ancestor of the forest-dwelling Pygmy of to-day, lower in mental level and more bestial in aspect than any of his [178] descendants, yet much advanced in mind beyond the man-ape of earlier ages, then we may with some assurance accept this as the type of the primitive man of Europe. He could have reached there by the land bridges which are thought to have connected Europe and Africa at that time, one closing the straits at Gibraltar, the other extending south from Italy by way of Sicily. These were the routes by which the apes are supposed to have entered Europe, and by which man may well have followed in a later age. It is possible, indeed, that man reached the northern continent from another locality, the habitat of the Negrito race in southeastern Asia and the Malaysian islands. The fossil man-ape of Java, Pithecanthropus, is a strong argument that this was the region, or one of the regions, in which the development of man took place. However this be, we can be assured that primitive man was far more likely to widen his field of occupation through migration than any other animal, and may conjecture that he spread over Europe and Asia in the mild preglacial times, and perhaps even reached America, giving rise to the early man of that hemisphere.

The advent of man in Europe was not probably followed by any considerable intellectual development. The mild and equable climate which at that time seems to have prevailed, was not likely to make a stringent demand on his mental resources. Food was very likely abundant and eas [179] ily obtained, animals of the chase being plentiful, and edible roots and fruits by no means lacking. Thus he could readily obtain the means of subsistence by aid of the arts and weapons employed by him in the tropical forests. It is not unlikely that some changes, both physical and mental, took place, but these were probably not great. There may have been some change in color and form, a first step toward the distinctions which separate the white from the black man, and a degree of mental adaptation to certain exigencies of the new situation; but in neither direction were the variations likely to be very decided.

Such, as we conceive it, was the man of early Europe, in great measure a counterpart of the forest nomad of the tropics of Africa and the East, the monarch of the animal kingdom, but not the lord of the earth. He may have made some progress in the contest with inanimate nature. Vegetable food in his new home was less abundant than in his old, and the instigation to agricultural pursuits was stronger. And though Europe was thickly wooded, it probably presented more open land than Africa. Both the incitement to agriculture and the facilities for its exercise were, in all probability, greater than in Africa, and man may have begun to cultivate the earth here at an earlier date than in his native realm. We are free at least to speculate that European man gained some slight knowledge of agriculture in the pre-glacial [180] period, but this is doubtful, and the relics of early man yield no evidence in its favor. Mentally it is questionable if he was advanced beyond the level of the least developed negro tribes, and perhaps not beyond that of the forest pygmies.

But at length the shadow of a mighty coming change began to fall upon the fair face of Europe. Year by year the winters grew colder. The ice sheet, which was in time to bury half of Europe under its chilly mantle, had begun its slow movement toward the south. It advanced very slowly. Centuries elapsed during its deliberate march. Had it moved with rapidity, few animals could have survived its effects. Some of them found time for changes in structure to fit themselves to the new conditions. Others perished as the wintry chill increased. Constituted for tropical warmth, they were unable to endure severe cold. The apes and monkeys may have been among the early victims. To-day the apes of Gibraltar are the only ones existing in a wild state in Europe, and it is doubtful if they are of an original stock. There is good reason to believe that escape by migration southward was cut off by the sinking of the ancient land bridges, so that the animals north of the Mediterranean had no choice between adaptation and annihilation.

Among the animals thus taken prisoner by the glacial chill was European man. He could not escape, and was forced to remain, exposed to the [181] alternatives of perishing from cold and hunger, or fitting himself to endure the new conditions which were coming upon his northern home, perhaps the most adverse to animal life that had ever been known. Man was about to be subjected to an extraordinary strain, which he could only meet by an extraordinary adaptation.

The changes by which he met these new conditions were in a very small degree physical; they were almost wholly mental. In all animals of the higher orders, adaptive variations are apt to be in a measure of this character, the body being relieved from the need of structural change through some new activity of the mind. In man this was undoubtedly the case in great, probably in very great, measure. There may have been an increase in size and strength, some variations in color, in the breathing organs, in power of resistance of the cuticle to cold, etc., but the principal physical change was in a growth of the brain and expansion of the cranium, giving rise to a less bestial physiognomy and an advanced mental power.

One physical change that would seem necessary to enable an animal to endure severe cold, the development of a thick protective covering of fur or hair, did not take place in man. The change was more likely in the other direction, since the hairy cover which is possessed by many of the forest folk has disappeared. This loss of hair by man has been referred by Darwin to sexual selection, that power [182] ful influence to which animals seem to owe so many physical structures of no apparent use, and some of them seemingly disadvantageous. In the case of man in the circumstances now under consideration, exposed without natural covering to the growing chill of the advancing ice sheet, the influence of sexual selection would certainly have found a strong counteracting force in natural selection, had not some other means of escaping the influence of the cold been found.

As it was, the difficulty was undoubtedly overcame in great measure by the adoption of artificial clothing. The mind came to the aid of the body. The man who could chip a stone into the shape of an axe or spear head, was sufficiently advanced mentally to conceive the idea of covering his body with leaves fastened together in some way, with other vegetable fabrics, or with the skins of slain animals. Protection from the cold was also sought in caverns and rock shelters, and for a very long period man remained a cave-dweller. There is hardly a cavern in western Europe in which he has not left some trace of his residence. Where caves were not available, rude artificial shelters were probably built. Even the orang builds a shelter of this kind, and we can readily conceive of man at a very early period making himself a shelter of leaves and boughs, from which, as the cold increased, he might easily evolve a hut composed of a wooden framework covered with skins such as he used for clothing.

[183] When and where the most important of discoveries, that of fire, was made, it is impossible to say. Fire arising from natural

causes, such as conflagrations started by lightning, no doubt early taught man the advantage of this agency as a protection from cold, but the artificial production of fire was a process too intricate to be arrived at by undeveloped man except as a result of accident. It has never been achieved, as we have seen, by the Andaman Mincopies. The rudiments of the fire-making art were possessed by primitive man. In chipping flints into arrow or lance heads sparks must frequently have been struck from the hard stone, and at times these may have fallen upon and kindled inflammable material. The rubbing requisite in shaping and polishing war clubs may have yielded a heat occasionally causing fire. In boring the holes necessary to make the needles found among primitive implements, a process resembling that of the fire-drill must have been employed. In short, it is not difficult to conceive of more than one way in which the fire-making art could have been gained by accident, though it may have been late in coming, since some, perhaps all, of the arts described were not attained until the Glacial Age. Once possessed, this important art would scarcely have been suffered to disappear. With its aid man could defy the effects of the glacial chill, so far as its direct action upon his body was concerned; and with it he also gained [184] a new and efficient means of defence against carnivorous animals, which have ever since feared fire more than weapons.

The discovery of methods of artificial fire-making was perhaps preceded by a utilization of the flames caused by lightning and other natural causes, the fire being conveyed by torches from hearth to hearth and kept alive with sedulous care. Even after artificial methods of fire-making were invented, our savage ancestors were exceedingly careful to keep their fires alive, as the Mincopies are today, and this heedful attention left its traces until very recent times. So important was the apparatus for kindling a flame deemed that in India the fire-twirl was made a god and became one of the chief deities of that polytheistic land. In many other places, especially in Persia, the element of flame was raised to the dignity of a deity and worshipped among the higher gods. Among the semi-civilized Americans the peril of the loss of fire gave rise to a serious religious ceremony. At certain set intervals all the fires within the limits of a tribe or nation were extinguished, and a period of gloom, despondency, and dread of the malignant powers succeeded. Then the "new fire" was kindled on the temple altar, and the flame was conveyed by swift messengers from hearth to hearth throughout the land. This done, the period of gloom was followed by one of general joy and festivity. The malignant deities were banished; the gods of light and warmth [185] were dominant again; happiness and security had returned to man.

The beginning of the use of clothing, of artificial shelter, and of fire formed one of the most vital periods in the history of human evolution. Coincident with them was the production of a much greater variety of implements than had been previously possessed, and many of these much superior to the older and ruder forms. The struggle with the glacial cold had roused man's mind out of its old sluggishness, and brought it actively into operation in devising means of counteracting the perils of his situation and fitting him to the new conditions of existence.

Among the important steps of progress was very likely a considerable advance in the use of language, enabling the men of that period more readily to consult with and advise one another, to give adequate warning of danger, to aid in the chase or in industrial pursuits, to educate the young and impart new ideas or teach new discoveries to the old. The mental powers of the best-trained individuals then as now served the whole community, and nothing of value that was once gained was likely to be lost. Discovery and invention at that early period probably went on with interminable slowness as compared with the progress in later ages, yet even then new ideas, one by one, came into men's minds, and step by step the methods of life were improved.

[186] One important effect of the glacial chill needs to be adverted to. The severity of the weather was not the only thing to be provided against. The discovery of fire and the invention of clothing and habitation were not enough to insure man's preservation. For the severe cold must have greatly changed the conditions of the food supply, and the man of the period found it a difficult matter to obtain the first necessaries of life. The easy-going man of the earlier age, living amid an abundance of fruits and vegetables and surrounded by numbers of game animals, or dwelling beside streams which were filled with easily taken fish, probably found the ques-

tion of subsistence one of minor importance. The coming on of the Glacial Age made this question one of major importance. The supply of fruits and vegetable substances was greatly decreased by the biting chill, and the number of food animals was correspondingly reduced; while through much of the year the effects of frost drove the fish from the streams, and cut off effectually this source of food. Man was brought into a situation in which only the most active exertion of his powers of thought could preserve him from annihilation.

He now found the exercise of the art of hunting more difficult than ever before, one that needed a new development of courage, cunning, alertness, and endurance, the scarcity of animals obliging him to make long journeys and attack the strong [187] est creatures. Whether or not he possessed the poisoned arrow, which the Pygmies now find so effective, cannot be said, but in all probability he was forced to invent new and more destructive weapons, a necessity that gave fresh exercise to his powers of invention. So far as our actual knowledge goes, the art of chipping stones into weapons and implements was not possessed before this period, and it may have been a result of the severe exigencies of the situation and the mental stimulation thence resulting. This art is not possessed by any of the Pygmies, the nearest approach to it being the splitting of stone by fire and using the splinters as weapons. Very likely preglacial man was similarly destitute of this art.

Under the severe strain of the glacial conditions the weak and incapable doubtless succumbed to the cold and deficiency of food; the strong and capable survived, gained superior powers, devised new weapons and implements, and became adapted to a new and decidedly adverse situation. From long depending, in considerable measure, on his physical powers, man came to trust more fully than before in his mental faculties, the result being a much greater variation in the size and activity of his brain than in other portions of his physical structure. While it had become more difficult to find and capture food animals, he was at the same time in greater danger from carnivorous beasts, which were forced by partial starvation to overcome their [188] dread of man. He was thus obliged to become as alert and ready in defence as he was in attack, to associate himself more fully with his fellows in his hunting excursions and his other labors, and to adapt the forms and forces of nature still more to his needs, his career as a tool-making animal being greatly stimulated by the necessities of his situation.

It is conceivable that the art of agriculture may have been one of the outcomes of the situation in which man now found himself. The decrease in the food supply must have put all his powers of invention to the test, and the probable diminution in number and productiveness of food plants may have served as an instigation to the cultivation of useful plants, and the preservation of their products, where possible, for winter supply. It is not unlikely that in this way and under this stimulation agriculture began, and that it made its way subsequently from this locality to more southern regions. In this, however, we cannot go beyond conjecture.

It seems useless to pursue this topic further, since the absence of facts forces us to confine ourselves largely to suggestions and probabilities. We have arrived at two definite hypotheses: first, that the original stage of man's progress upward from the apes was completed when he gained dominion over the animal kingdom and attained the condition of the forest pygmies; second, that an advanced stage was reached when he achieved the conquest [189] of nature, so far as overcoming the exceedingly adverse conditions of the Glacial Age was concerned. At the close of this period of frigid cold man emerged as a higher being than the forest nomad or the agricultural people of the tropics, possessed of much superior arts and implements and with largely enhanced mental powers. The long and bitter struggle for existence through which he had passed had lifted him to a much higher level in the upward progress of life.

He was a savage still, and at the close of the struggle he settled down into a second stage of stagnation. The conflict was at an end, he was the victor in the fight, he could rest upon his laurels and take life easy. In addition to his mechanical gains, man had advanced much in social and political relations, and continued to advance until his primitive form of organization was perfected. At the end of it all we find him existing under two conditions, depending upon differences in the character of the country in which he lived.

In the steppes and deserts of Asia and the deserts of Africa he was a nomad herdsman, his life being spent in the care of his flocks and herds, his political organization the patriarchal, his possessions few, his needs small, his mind at rest, his progress largely at an end. Thus he still lives, and this organization and mode of life still persist, little affected by the long centuries that have passed and not greatly modified by the many wars in which [190] he has been engaged. Mentally, the man of the steppe and the desert is to-day little advanced beyond his predecessors of thousands of years ago.

In the more fertile regions of the earth man had become an agriculturist, each clan holding its section of the earth as common property. A different though primitive form of political organization arose here, that of the village community, in which there was no distinction of rich and poor, all men were equal in rights and privileges, all were content with their situation, and the mental condition was largely that of stagnation. This political condition we find to have been widespread over the earth, alike in the eastern and western hemispheres, as the one into which all developing agricultural communities emerged, and in which they persisted unchanged until forced to adopt new relations through a new influence still to be described. As the patriarchal clan is persistent on the Asiatic steppes and deserts, so is the village community on the Russian plains and among the Aryans of Hindostan. It has been generally overcome in other localities, but it was broadly extended until within comparatively recent times, and traces of it may still be found in many parts of the earth.

The political organization of these primitive communities of herders and farmers was of the simplest. Over the herding clan a patriarchal chief presided, his authority based on his position [191] as representative of the ancestor of the community. The head man of the agricultural clan was elected by the free choice of his fellows, his equals in rank and station. But the supposed most direct descendant from the clan ancestor was apt to be chosen. In both cases the political organization was of the family type, being but an extension of family government, and the widely prevailing system of ancestor worship had much to do with the reverence in which the chief was held and the authority which he exercised.

The development of this phase of human progress did not stop here. Kingdoms and empires arose as direct resultants of this condition of affairs. In some localities, such as Egypt and Babylonia, the great fertility of the soil in the time gave rise to a dense population, largely gathered in towns and villages, where industries other than agriculture developed and closer social relations existed. The simple organization of the village or the clan was not sufficient for such a population, and a more intricate governmental system arose; but it seems to have been simply an extension of the older system of chieftainship, based on the family or paternal relation, and on the growth of religious influence and priestly control. It seems, in fact, to have been through the influence of religious ideas that men first rose to power and became supreme over their fellows.

We have no concern here with the development [192] of religious systems, other than to say that in the primitive agricultural community a succession of ideas of man's relation to the unseen arose, yielding, in addition to the widespread ancestor worship, a system of shamanism, or belief in the presence and power of malignant spirits, and one of fetichism, which developed into mythology, or worship of the great powers of nature. What we are concerned in is the fact that from these religious conceptions a priesthood everywhere arose, beginning in the simple conjurer or the healer by spells and incantations, and developing into a priestly establishment whose leading members had a vigorous control over the people through their beliefs, fears, and superstitions.

This priestly system was the basis of the first imperial organization. Kingly authority was not gained at first through power over men's bodies, but through influence over their minds. There is much reason to believe that the chief of the clan or tribe, who led in its public worship and was looked upon as the representative of its divine ancestor, retained the influence thence arising as the tribe developed into the nation, adding the power and position of the high priest to that of the tribal chief.

There is abundant evidence that in this simple and direct manner the imperial organization everywhere grew out of the primitive village and patriarchal systems. In the early days of Egypt, before [193] its era of conquest began, the Pharaoh was the high priest of the nation, weak in temporal, strong in spiritual power; and the political organization in general probably grew out of the sacerdotal establishment. Very likely the Babylonian kingdom was organized in the same manner, though wars and changes of dynasty have obscured its early state. In China the patriarch of a nomad horde became emperor of a nation retaining ancestor worship as its chief religious system. He held, and still holds, the position of father of his people, the representative of the original ancestor, and high priest of the nation.

In India the priestly establishment was differently organized. It was a democracy instead of an aristocracy. There was no high priest to seize the reins of government. As a result, no empire arose in India. A simple outgrowth of the tribal system developed, each tribe under its chief, while the priesthood as a whole remained the real rulers of the people.

If we come to America, we discover a similar condition of affairs, the head of the religious establishment becoming everywhere the head of the nation. This was the case in Mexico, where the Montezuma was high priest, and derived his power largely from this position. It was the case in Peru, where the Inca was the direct representative on earth of the solar deity. It was the case with the agricultural communities of the southern United [194] States, whose Mico was at once high priest and autocrat. It was doubtless the case with the Mound Builders, of whom these communities were probably the descendants.

Such seems to have been the final outcome of the contest with nature, where permitted to develop in its natural and unobstructed way. A series of empires of a simple type of organization arose, their rulers uniting temporal and spiritual power, and becoming autocrats in a double sense, supreme lords of body and soul. It was in its nature a persistent type. Once reached, it tended to continue indefinitely, stagnation following the era of growth. But war and invasion have broken it up everywhere except in China, a country largely defended by nature against invasion and inhabited by an innately peaceful people. As the forest Pygmy group represents today the completion of the first stage of human evolution, so the patriarchal empire of China represents that of the second. Stagnation there long since succeeded development. For several thousand years China has almost stood still. It comes down to us as the fossilized representative of an antique system, physically active but mentally inert, its organization rigidly fixed, and not to be disturbed unless the empire itself is rent to pieces.

[195]

XI

WARFARE AND CIVILIZATION

Long before the second phase of the evolution of man had been completed the third phase had begun, that of the conflict of man with man. The animal kingdom once subdued, and nature made man's friend and servant, the human race increased and multiplied until the borders of communities met and hostile relations arose between them. A fight for place began, a struggle for dominion, a fierce and incessant contest for supremacy, and for ages men locked arms in a terrible and merciless strife, in which the weak and incompetent steadily went to the wall, the strong, daring, and aggressive rose to power and control.

It was the final act in the great drama of "natural selection," which had been played upon the stage of the earth since the first appearance of living forms; the last and most ruthless of them all, for the instigating cause was no longer merely the pressure for a share of the food supply, but to this was added the lust for power and place, the hunger for wealth and dominion, the insatiable appetite for autocratic control. Millions upon millions of men were swept away by the sword, and [196] by its attendant demons, famine and pestilence; and still the stronger and abler climbed to the top, the weaker and inferior succumbed; and the intellectual evolution of man went on with enhanced rapidity as the harvest of the sword was gathered in, and the merciless reapers of men swept in successive columns over the earth, each a stage higher in mental ability than the preceding.

This phase of human evolution is that of the era of human history. Before its advent man had no history. It would be as useful to attempt to give the history of the gorilla as of man in the early stages of his progress. History is the record of individuality, and in primitive times equality and communism prevailed, and the individual had not yet separated himself from the mass. Man had settled into the dull inertness of a stagnant pool, and the fierce winds of war were needed to break up his mental slothfulness and stir thought into healthful activity. There must be leaders before there can be history; the annals of mankind begin in hero worship; the relations of superior and inferior need to be established; and individual action and supremacy are the foundations upon which all history is built. Only by stirring up the deep pool of human life into seething turmoil and unrest could the tendency to stagnation be overcome, the best and most aspiring rising to the top, the dull and heavy sinking to the bottom, and the element of thought permeating the whole with its vitalizing spirit.

[197] When this phase of evolution is reached, we cease for the first time to deal with species and genera in the mass and begin to deal with individuals, who now emerge from the general group and stand above and apart like great signal posts on the highway of progress. These heroes are not alone those of the sword. They are the leaders in art, in literature, in science, in thought, in every domain; the men who stand, above, supreme and shining, and toward whose elevation the whole mass below surges slowly but strenuously upward. The third phase of human evolution, therefore, is that of the emergence of the individual as the leader, lawgiver, teacher of mankind, each leader forming a goal for the emulation of all below. And this condition is the legitimate outcome of war, which, terrible as it always has been, was the only agency that could rapidly break up the stagnancy of early communism and send man upward in a swirl toward the heights of civilization.

To give the history of this phase of evolution would be to give the history of mankind, and would be aside from the purpose of this work. All that need be attempted, in support of our argument, is to present some general deductions from human history, indicating the leading features of the service man has received from war.

Conflict between man and man was at first vague and inconsequential. It was only after settled and organized communities, based origi [198] nally on the family relation, and held together by the possession of property in common, had been formed, that war became more effective in its results. The chief of these results, in the early days, were two: the breaking up of the old equality of power and possession, and the development of larger and more powerful communities. The head man of the village community, or the herding clan, possessed some delegated authority but no political supremacy over his fellows. Equality existed alike in theory and in fact. Battle between neighboring clans was the first step toward breaking this up. The conquered clan became subordinated to the victorious one, and the chief of the victors, as the representative of his clan, exercised an authority over the subject community which he did not possess at home. The degree of subordination differed from the mild form of tribute-paying to that of personal slavery. But in either case we see the old condition of equality vanishing, and that of class distinctions and the relation of superior and inferior arising, while the power of the chief advances from a delegated authority to an established supremacy.

The second outcome of this early phase of war was an increase in the size of political groups. The conquered were forced to aid the conquerors in war as in peace; clans combined to resist aggression; minor communities grew into organized tribes; tribes developed into nations as a result of [199] warlike operations. This growth in political organization was a necessary and inevitable result of continued warfare. The aggressors gathered all the strength possible. The assailed peoples did the same. Temporary alliances grew into permanent ones. Larger armies were formed, larger communities were organized, national development advanced at a rate tenfold, probably a hundredfold, more rapidly than it would have done had peaceful conditions persisted.

Side by side with tribal and national consolidation went on the growth in leadership. The head man became a war chief, the war chief a king. Success made him a hero to his people. He grew to be the lord of conquered tribes; into his hands fell the bulk of the spoils; the relation of equality of possessions vanished as the plunder taken by the army was distributed unequally among the victors. Below the principal leader came his ablest followers, each claiming and receiving a proportionate share in the new division of power and wealth. In short, when the era of war had become fully inaugurated, the old social and political relations of mankind were broken up with great rapidity; equality of power was replaced by inequality, which steadily grew more and more declared; equality of wealth in like manner vanished; in all directions the individual emerged from the mass, class distinctions became intricate, and the relations of rich and poor, of king, noble, [200] citizen, and slave, completely replaced the old communal organization of mankind.

War was the great agent in this evolution. It might have emerged slowly in peace; it came with almost startling rapidity in war, and reached a degree of power on the one hand and subordination on the other that could scarcely ever have appeared had conditions of peace prevailed. With this growth of great nations came a rapid development in political science, in legal institutions, in social relations. An enormous advance was made, in a limited period, in the civilization of mankind; as a result, not of the devastation and slaughter of war, but of its influence upon human organization.

It was the principle of reward for ability to which the leaders of men owed their supremacy. When nations were organized this same principle took another and very useful form. The distribution of wealth had become strikingly unequal. There were endless grades of distinction between the supremely wealthy and the absolutely poor. The wealthy were ready to lavish their money in return for articles of pleasure and luxury. The poor, in their thirst for a share of wealth, were strongly stimulated to inventive activity in producing new and desirable wares. Inequality became the mainspring of business activity; thought and inventive ingenuity were strongly exercised; a rapid progress went on in the production of new devices, new methods, and new articles of necessity and lux [201] ury; manufacture flourished, commerce increased, civilization appeared, the whole as a legitimate outcome of the conditions brought about by war.

This phase of human evolution, as may be seen, was radically different from that already considered, arising from the development of sacerdotal influence and priestly power. They worked together, no doubt. The establishment of the great primitive empires, as a peaceful process, was greatly complicated by war, which tended steadily to increase the temporal power of the ruler and enable him in time to control by the sword alone. But it is interesting to find that long after the old system was practically overthrown its shadow still lay upon the nations. The powerful war monarchs of Assyria led their armies to conquest in the name of the national deity, whose vicegerents they claimed to be. The autocratic emperors of Rome went so far as to claim in some cases to be gods themselves. Even in modern Russia some of this dignity pertains to the emperor, as the supreme head of the national church. Old ideas are proverbially hard to kill.

But the mission of the priesthood by no means stopped here. The priests rose to influence as the teachers as well as the leaders of the people. The members of this class, set aside from manual occupations, and devoted to thought upon the relations of man to the divine, played an important part in the development of the human mind. As a result [202] of their speculative activity of thought the old religious systems sank into the background; the simple worship of primitive times was overshadowed by intricate mythological systems, splendid in worship and creed; cosmogonies and philosophies were devised; and human thought, once fairly set loose in this field, went on with great energy and imaginative fervor.

Literature arose as a result of this activity of thought. It took at first the form of hymns, speculative essays, magical formulas, dogmas, ordinances of worship, etc. By degrees it grew more secular in form, until in the end secular literature arose. This was greatly stimulated by the conditions of inequality arising from war. In the same manner as the reward for merit in invention stimulated men to activity in the mechanical arts, so the hope of reward for literary production stirred up men to the composing of poems, histories, and other works of thought. In both directions, physical and mental, men were stimulated to the most active exertions by the conditions of inequality in wealth and power, and the consequent desire to obtain a share of the money lavished by the rich and the authority similarly lavished by the powerful.

The broad general view here taken must suffice for our consideration of this phase of human evolution. It brings the story of the development of man closely up to the present stage of political [203] and social organizations and relations. It may be said, in conclusion of this section of our work, that the powerful agency of war, so active and important in the past, has in great part lost its utility in the present, and bids fair to be brought to an end before the world is much older. It is no longer needed, nearly or quite all that it is capable of doing for mankind being accomplished, while the equally powerful agencies of commerce, travel, leagues of nations, and other conditions of modern origin have taken its place.

War, while yielding many useful results, has given rise to others whose utility is questionable, and whose ill-effects it will take much time and effort to set aside. The inequality of power to which war gave rise continues in many parts of the world, and the inequality of wealth shows signs of increase instead of diminution. Once useful, they have developed to an injurious extent. The result is a state of unrest, discontent, and more or less active opposition, which constitutes a condition of permanent conflict, a deep dissatisfaction with existing institutions abnormal to a justly organized society. War has become in great measure useless; but the scaffolding from which it built up the edifice of civilization remains, and stands as a tottering ruin threatening to engulf mankind in its fall.

Ever since the triumph of autocracy in the Roman empire, the masses of mankind have steadily protested against an inequality that is [204] alien to the natural rights of man. For century after century the struggle against undue exercise of power has gone on, and the hereditary lords of mankind have lost, stage by stage, their usurped power, until in the modern republic they have been replaced by the servants and chosen agents of the people. But the autocracy of wealth still holds its own, and is growing more and more formidable, and against this the wave of opposition is now rising. Everywhere man is earnestly and sternly demanding an equitable distribution of the productions of nature and art. What the outcome of this demand will be it is impossible to say. It must inevitably lead to some readjustment of the wealth of mankind; but only the slow process of social evolution can decide what this shall be.

We have endeavored in this brief treatise to trace the development of man from his primeval state as a tree-dwelling animal in the depths of the tropic woods, through the phases of his later condition as an erect surface dweller, his conflict with and dominion over the animal kingdom, his subsequent contest with the adverse powers of nature, and his final warfare with his fellows and emergence into civilization. Each of these contests has left its results; the first in the forest nomads of the eastern tropics, the second in the patriarchal herding tribes of the steppes and deserts, the village communities of Russia and the paternal empire of China, the third in the enlightened nations of Europe and America.

[205] For how long a period this mighty drama of evolution has continued it is impossible to say. Its first phase must have been of interminable slowness; its second, while more rapid, still very deliberate; its third of much greater rapidity, yet extending over several thousands of years. Millions of years have probably passed away since it began, yet the period involved is none too long for the magnitude of the results, whose greatness can be seen if we contrast man's mental development with that of the lower animals during this period. Physically, the development of man has been inconsiderable — much less apparently than that of many other animals. Mentally, it has been enormous. The whole of nature's influences, in new and often adverse situations, have been brought to bear upon man's mind, and as the result we have civilized man as contrasted with the anthropoid ape. And the end is not yet. The era of war in man's development is near its close, and a new era of peace, under conditions of advanced mental and physical activity, seems about to begin. Its outcome no man can predict, but it may far surpass in beneficial results all that has gone before, and carry man upward to an extraordinarily elevated mental plane.

[206]

XII

THE EVOLUTION OF MORALITY

The evolution of man from his animal ancestry has been a composite phenomenon, one by no means confined to the physical and intellectual conditions which we have so far considered, but embracing also features of moral and spiritual progress. The origin and growth of these need also to be reviewed, if we would present a fully rounded sketch of human evolution. So far as his physical form is concerned, man became practically completed ages ago, as the supreme effort of nature in the moulding and vitalizing of matter. When the arena of the struggle for existence became transferred from the body to the mind, variation in the body, once so active, rapidly declined; and with the full employment of the intellect in the conflict with nature, physical evolution ceased, except in minor particulars, and the organic structure of man became practically fixed. The human animal, therefore, as a physical species, has reached a stage of permanence. And this may be regarded as the supreme result of material evolution in animals; or at least it may be affirmed that, while man continues to exist, no member of [207] the lower animal tribes can possibly develop to become his rival.

But though man is not markedly distinct as a physical species from his anthropoid ancestor, the process of evolution has not ceased, but has gone on in him rapidly and immensely. The strain has simply been transferred from the body to the mind, and to the extent that the mental characteristics are more flexible and yield more readily to formative influences, the mind has surpassed the body in rapidity of evolutionary variation. Within a period during which the lower animals have remained almost unchanged, man has varied enormously in mental conditions, and to-day may be looked upon, not merely as a distinct species, but practically as a new order, or class, of animals, as far removed intellectually from the mammals below him as they are from the insects or mollusks.

If now we turn from the physical and intellectual to the ethical stage of development, it will be to perceive as marked and decided a process of evolution. The change has, perhaps, been even greater, since in the lower animals the moral faculties are more rudimentary than the intellectual. But, on the other hand, the moral development in man has been much inferior to the intellectual. Therefore, though the foundation was lower, the edifice has not reached nearly so great a height, and man to-day stands in moral elevation considerably below his intellectual level.

[208] It was formerly the custom to look upon man as the only intellectual and moral animal, the forms below him being credited solely with hereditary instincts. This belief is no longer entertained by those familiar with the results of modern research. Evidences of unquestionable powers of thought have been traced in the lower animals, imagination and reason being alike indicated. The elephant, for instance, is evidently a thinking animal, and is capable of overcoming difficulties and adapting itself to new situations, using methods not unlike those which man himself might display under similar circumstances. Its gratitude for favors and remembrance of and revenge for injuries are evidences of its possession of the moral attributes. The recorded instances of displays of reason in the dog, man's constant companion, are innumerable. Intellectual attributes are still more pronounced in the ape tribe, as indicated in a preceding chapter, where it was argued that man began his development in intellect at a somewhat advanced stage.

The same cannot be said in regard to his moral evolution. In this respect the level from which man emerged was a much lower one. If his moral growth may be symbolized as a great tree, it is one not very deeply rooted in the world below him. Yet it doubtless has grown out of the soil of animal life, and its finer tendrils and fibres may be traced to a considerable depth in this fertile soil.

[209] Before proceeding with this subject, it is important to devote some attention to the characteristics of the moral attributes, concerning which there is much diversity of opinion. There has been abundance of theorizing upon the principles of ethics, thinkers dividing themselves into two widely separated groups. In the one school, the intuitive, the principles of morality are looked upon as inherent in the soul of man, unfolding as the plant unfolds from its seed. In the other school, the inductive, morality is claimed to be founded upon selfishness, the moving principle of human actions being the desire to avoid pain and attain pleasure. Each school makes a strong ar-

gument, which goes far to indicate that each is based upon a truth, and therefore that neither has the whole truth.

The fault would appear to lie in the attempt to make morality a unit. In our view this unity does not exist. While both schools may be partly right, neither would seem to be wholly right, and they appear to be pulling at the two ends of a single chain. Ethics, in short, may be regarded as composed of unlike halves, which unite centrally to form a whole. It may aid to reconcile the conflicting systems of theorists if it be held that the inductive half of ethics is the product of the reasoning powers and outer experience, the intuitive half the product of feeling and inner development; while both meet and harmonize in life as reason and feeling harmonize in the mind.

[210] It is interesting to find that it is the intuitive, not the inductive, element of the moral attributes that we find principally developed in the lower animals. This is the outgrowth of instinct, not of thought; the development of that principle of attraction which manifests itself in all nature, and which, when associated with consciousness, becomes what we know as love, affection, or sympathy. It is a powerful and pervading force in all matter, intelligent and unintelligent, and in conscious beings falls naturally among the emotions. Like all the passions, it is instinctive in origin, though it may come under the control of the intellect as the mind develops. In the lower animal world it is manifested as a vigorous attraction, the sexual. In the higher animals this attraction expands and grows complex. The attraction between the sexes becomes love, and in its full unfoldment may join two individuals together for life and influence most of their actions. To the attraction between the sexes should be added that between parents and children, the parental and filial, and that between associates, the tribal or social, the latter, though weaker, of the same character.

With these bonds reason has nothing to do. It does not form them and would seek in vain to sever them. They belong to a part of the mental constitution which lies outside the kingdom of thought, and they, therefore, often act counter to the selfish consideration of personal safety. The love bond, [211] indeed, in its full strength, seems to constitute a partial loss of individuality. Mates will suffer pain and endure physical injury for each other or for their offspring to as great an extent as if these constituted a part of themselves, and as if their actions were performed in self-defence.

With this brief review of the philosophy of the ethical sentiments, we may proceed to a consideration of the facts. While the rudimentary form of the sentiment in question is manifest far down in the descending grades of animal life, it expands into what we may fairly term love or affection only in the higher forms. Romanes, in his "Animal Intelligence," remarks: "As regards emotions, it is among birds that we first meet with a conspicuous advance in the tenderer feelings of affection and sympathy. Those relating to the sexes and the care of progeny are in this class proverbial for their intensity, offering, in fact, a favorite type for the poet and moralist. The pining of the 'love-bird' for its absent mate, and the keen distress of a hen on losing her chickens, furnish abundant evidence of vivid feelings of the kind in question. Even the stupid-looking ostrich has heart enough to die for love, as was the case with a male in the Rotund of the Jardin des Plantes, who, having lost his mate, pined rapidly away."

Among social and communal animals the sentiment of sympathy widens to embrace all the members of the tribe, a characteristic which is very [212] strongly manifested in so low an organism as the ant. As an example of this feeling among birds, Romanes quotes an interesting illustration from Edward, the naturalist. The latter had shot and wounded a tern, but before he could reach it, the helpless bird was carried off by its companions. Two of these took hold of it by the wings and flew with it several yards over the water. They then relinquished their burden to two others, and the process continued in this way until they at length reached a rock at some distance. When the hunter, eager for his prize, pursued them, the sympathetic birds again took up their wounded companion and flew off with it again over the water.

Abundant instances of this sentiment of social affection could be quoted from the mammalia. It is by no means confined to members of a species, but may extend to very unlike species. No one needs to be told of the warm affection so often shown by the dog for its master, a love which will lead it to dare wounds or death in his service, or in the protection of his property. This altruistic sentiment strongly exists in the monkeys. Examples of the ardent feeling of these animals for their fellows have been given in a preceding chapter, and many more might be quoted, if necessary. It must suffice here to quote a single further instance cited by Romanes, and relating to a small monkey which was taken ill on shipboard, where there were several others of different species.

[213] "It had always been a favorite with the other monkeys, who seemed to regard it as the last born and the pet of the family; and they granted it many indulgences which they seldom conceded to one another. It was very tractable and gentle in its temper, and never took advantage of the partiality shown to it. From the moment it was taken ill, their attention and care of it redoubled; and it was truly affecting and interesting to see with what anxiety and tenderness they tended and nursed the little creature. A struggle often ensued between them for priority in these offices of affection; and some would steal one thing and some another, which they would carry to it untasted, however tempting it might be to their own palates. They would take it up gently in their forepaws, hug it to their breasts, and cry over it as a fond mother would over her suffering child."

With the human race the love sentiment does not usually display the singleness of energy shown among the lower animals. It is affected and often checked in its development by an intricate series of influences, which act on savage and civilized man alike. The family formed the primitive human group, its linking elements being the sexual attraction between man and woman and the fervent affection between parents and children. These feelings, while strong in certain directions, were crude and uneven. In savage tribes to-day the wife is an ill-treated drudge. Yet the husband will protect his [214] wife and children from danger at risk of his life. The maternal instinct seems still stronger. The mother often acts as if the child were an actual part of herself. Danger or injury to it produces in her a mental agony, the close equivalent of its fear or pain, and she will endure suffering and peril in its protection with an impulse beyond the control of reason.

This sentiment, in a weakened form, extended from the family to the group; and the success of man in gaining the mastery over the other animals was doubtless greatly aided by the strong bond of social affinity existing between the members of a group. They worked together in a fuller sense than any other animals except the ants and bees.

From the original social group another and closer community seems gradually to have developed, the group of kindred. This was a natural outgrowth from the family, whose bond of affection was extended to include more distant relatives, until there emerged the organized group of kindred known as the "Village Community," which seems everywhere to have preceded civilization. This bond of kindred gradually extended, combining men into larger and larger groups, until the clan, the horde, and the tribe emerged, their members all linked together by the reality or the fiction of common descent. Such was the form of organization that existed in Greece and Rome in their [215] early days, and made its influence felt far down into their later history. It existed indeed, at some period, over almost all the earth.

As the group widened, the bond of sympathy weakened. Love in the family found its counterpart in fellow-feeling in the tribe, in patriotism in the nation. It is undoubtedly true that desire for personal protection is one of the strong influences which bind men into societies. The hope of advantage in other directions and the pleasure of social intercourse are other combining forces. Yet below these rational elements has always abided the emotional element, the sympathetic attraction which binds kindred closely together, and which exerts some degree of influence on all members of the same group or nation.

The development of the ethical principle in mankind is largely due to the extension of the sentiment of social sympathy. For ages it was confined to the immediate group. Such was the case even in civilized Greece, intellectually one of the most advanced of peoples, but morally very contracted. The Greeks were long divided into minor groups, with the warmest sentiment of patriotism uniting the members of each community, while their common origin bound all the Hellenes together. But this feeling failed to cross the borders of the narrow peninsula of Greece, all peoples beyond these borders being viewed as barbarians, in whose pleasures and pains no interest was felt, and whose misfortunes [216] produced no stir of sympathy in the Grecian heart. Even Aristotle taught that Greeks owed no more duties to barbarians than to wild beasts, and a philosopher who declared that his affection extended to the whole people of Greece was thought to be remarkably sympathetic.

The Romans were equally narrow in their early days, and not until the empire extended to the outer borders of the civilized world did this narrowness give way to a more expanded sympathy. The brotherhood of mankind, indeed, was taught by Socrates, Cicero, and others of the ancient moral philosophers, yet these seeds of philosophy fell in very sterile soil and took root with discouraging slowness. Philosophers elsewhere taught the dogma of universal love, — Confucius among the Chinese, Gautama among the Hindoos, — but their teachings have borne little fruit in the great, stagnant peoples of Asia, in whom the narrowness of semicivilization prevails.

The teachings of Christ, whose code of morality was the intuitive one, "Love one another," have been far more effective. Christianity became the religion of Europe, since then the most progressive part of the world, and with every step of progress in civilization the Christ doctrine of charity and sympathy reached a higher and broader stage. To-day it has attained, in Europe and America, a wide degree of development, and the vast extension of human intercourse through the mediums of [217] travel, commerce, and telegraphic communication is, for the first time in human history, beginning to lift the doctrine of the universal brotherhood of man from the plane of a philosophic dogma toward that of an established fact. The range of sympathy is narrow yet, selfishness predominates, the truly altruistic are the few, the feebly sympathetic and coldly selfish are the many; yet it must be admitted that there has been a great development of altruism during the nineteenth century, and the promise of the coming of Christ's kingdom on the earth is greater to-day than at any former period in the history of mankind.

The love principle is the innate moral element of the universe. Its rudimentary form is the attraction between atoms, which expands into the attraction between spheres. We see a development of it in the magnetic and electric attractions, and a higher one in the sexual attraction that exists in the lowest organisms. Its expansion continues until it reaches the high level of human love and social sympathy. But throughout its whole development consciousness takes no part in its origin. While conscious of its existence, we do not consciously call it into existence. Men and women "fall in love"; they do not reason themselves into affection. Those we love become in a measure a part of ourselves, we feel their sufferings and endure their afflictions, not through the nerves of the body, but through the finer ones of the mind, [218] —a plexus of spiritual nerves which stretch unseen from soul to soul. So strong is this sympathetic affinity that Comte was induced to look upon mankind as an organism, and it gave rise in the mind of Leslie Stephens to the conception of a common "social tissue."

Love and law rule the universe. It is this second moral element, that of law, which we have next to consider. Inductive morality had its origin in experience; it assumed the form of social restriction, then of fixed law and precept, and culminated in the sense of duty—a conscientious avoidance of that which was thought to be wrong, and an earnest desire to do what was looked upon as right.

The history of this phase of morality differs essentially from that of the phase we have just considered. The sense of duty, the conscientious sentiment, so highly developed in man, seems largely non-existent in the lower animals, so far as observation has taught us. Yet it is not quite wanting, its rudiment is there, and this rudiment is capable of development. It may be, indeed, that a highly developed sense of duty exists in the ants and bees, to judge from their diligent labors for the benefit of the community. But the clearest examples of conscientious performance of duty are those seen in the case of the dog, in which animal intimate association with man has developed something strongly approaching a conscience. A dog needs only to be well treated to [219] display a sense of dignity and a self-respect analogous to these feelings in man. A sensitive resentment against injustice in high-caste and carefully nurtured dogs has often been observed; while shame for an act which the animal knows to be forbidden has been seen in a hundred instances. The sense of duty is occasionally very strongly developed. Many striking examples of this are on record. A dog will often defend his master's property with the greatest devotion, letting no temptation draw him away from the path of duty.

An instance has been related to the writer in which an extraordinary display of this feeling was made. A gentleman, on coming home at night, found he had forgotten his key, and attempted to enter the house by the window of a room in which his dog was on duty as a night-watch. To his surprise and annoyance the animal would not permit him to enter, and attacked him every time he tried to climb in. The animal knew him well, responded to his attempts to fondle it, but the moment he made an attempt to enter the window it became hostile and seemed ready to spring upon him. In its small brain was the feeling that no one, master or stranger, had the right to enter that house at night by the window, and it was there to perform its duty without regard to persons. In the end, the gentleman was obliged to leave and seek shelter elsewhere.

The development of the sense of duty and [220] the growth of moral restriction in primitive man were probably very slow, much more so than the evolution of intelligence. The social habit of man doubtless rendered necessary, at an early period, some restraints on the actions of individuals, and these in time gained the strength of unwritten law; but many of them were scarcely what we should call moral obligations. Many such restrictions exist among savage tribes to-day, and to these we must turn for examples of their character. We, for instance, look upon theft and lying as immoral practices, but such is not the case with savages generally, most of whom will steal if the opportunity offers, while they will lie in so transparent and useless a manner as to indicate that they see nothing wrong in this practice. And yet the aborigines of India, many of whom are very immoral according to our standard, are often strongly averse to untruthfulness. "A true Gond," says Mr. Grant, "will commit a murder, but he will not tell a lie." It is well known that truthfulness was one of the chief virtues of the ancient Persians, a virtue that was accompanied by much which we would call immoral. The Hindoo devotee is exceedingly tender of the lives of animals, while he is often callous to human suffering. Disregard of human suffering, indeed, showed itself strongly through all the past ages, men being slaughtered with as little compunction as if they were so many wild beasts, while fright [221] ful tortures were inflicted with an extraordinary absence of humane feeling. And these excesses were committed by persons who in the ordinary affairs of life were frequently tender in feeling and conscientious in action.

In truth, moral development from this point of view has always shown a one-sidedness that goes far to discredit the doctrine of intuitive conceptions of right and wrong. The indications are strong that rules of conduct are not inherent in the human mind, that men become moral to the extent that they are taught the principles of justice, and grow one-sided in their ideas of virtue through incompleteness in their moral education. What we call sinfulness is largely a matter of custom and convention. Men cannot properly be said to sin when their actions are checked by no conscientious scruples, and what one people would consider atrocious instances of wrongdoing, might be looked upon as innocent and even estimable by a people with a different moral standard. Religion has much to do with this. The human sacrifices and cannibal feasts of the Aztec Indians, for instance, were regarded by them as good deeds, obligations which they owed to their gods. Yet this people had attained to some of the refined practices and moral ideas of civilization.

The leading principles of correct human conduct are few and simple. They were arrived at early in the history of human thought, and little has since [222] been added to them. They arose as results of human experience, as necessary principles of restraint in developing communities, and were nearly all extant in prehistoric times as the unwritten laws of social organization. What creed-makers did was to put these ancient axioms of morality on record, and offer them to the world as codes of religious observance. They could not have been of primitive origin, since the most of them do not exist among the savage tribes still with us. There is nothing, indeed, to show that any idea of sinfulness exists in the minds of the lowest savages, the rules of conduct which they possess being such regulations as are necessary to the existence of the most undeveloped community.

Of the various codes of morals, much the best known to us is that given to the Israelites by Moses, the famous "Ten Commandments." The most of these—as of all such codes—were evidently legal in origin, rules necessary for the existence of a civilized society, restrictions controlling the conduct of men toward one another. It was the creed-makers who first gave such legal restrictions the strength of moral obligations, and announced that their infraction would be punished by divine agencies, even if they should escape human retribution.

Many hurtful acts, indeed, came to be viewed as crimes alike against God and man, and punishable in the interests of both. Political and moral obli [223] gations thus shaded together; some of the evils of the world being punished by human agencies alone, some by divine, some by both. It must be said, however, that throughout the whole progress of human civilization the influence of moral obligations has been rising, while the necessity for political laws has declined in like proportion. In ancient times the penalties for crimes against the community were terribly severe, while religion threatened those who offended the divine powers with frightful future punishments. The necessity for such severe restrictions has long been decreasing, and the more vividly it is felt that immoral deeds or debased thoughts and purposes will be visited by a spiritual retribution, the less necessity is there for laws and penalties. Thus the limitation of human actions by government is growing less necessary than of old, in conformity with the growing sense of spiritual degradation in evil and of spiritual elevation in good deeds. Mild laws have succeeded the severe edicts of the past, and with a considerable section of the community restrictive laws have become useless, conscience taking the place of law. In such men the impulse to evil deeds dies unfulfilled, and the penalty for wrong-doing within themselves may be more severe than that which the community would inflict. In the souls of such men sits a spiritual tribunal by which evil thoughts are tried and punished before they can develop into evil acts.

[224] This consideration of the development of the moral principles and dogmas has been necessarily brief. In what direction it is leading must be evident to all, and we can with assurance look forward to a condition of human society in which conscience will have

become a stronger element of the intellect than now, the sense of moral obligation a more prevailing sentiment, and legal restriction a less necessary governmental requirement.

Of all the isms of the day altruism is far the noblest and most promising. In this opponent of selfism, this regard for the rights and happiness of others equally with our own, we find the link which binds together the two halves of the moral principle. The love sentiment on the one hand, the sense of duty on the other, meet and combine in the zeal of altruism, for which a truly developed conscience is merely another term. Those who have the good of others strongly at heart, who are truly Christian in a practical realization of the brotherhood of mankind, can safely be set free from all the reins of law, and trusted to do the right thing from innate feeling instead of outside compulsion. And, trusting in the future full development of the altruistic sentiment, we can hopefully look forward to a time in which the moral law will exist alone, conscience become the controlling force in human actions, and government let fall the whip which it has so long held in threat over the shrinking back of man.

[225]

XIII

MAN'S RELATION TO THE SPIRITUAL

The purpose of this work has been to trace the evolutionary origin of man, in his ascent from the lower animal world to his full stature as the physical and intellectual monarch of the kingdom of life. But to round up the story of human evolution it seemed necessary to consider man from the moral standpoint, and it now appears equally desirable to review his relations to the spiritual element of the universe. Having dealt with the development of man as a mortal being, we have now to regard him as a possibly immortal being.

This outlook into the supreme domain of nature lifts us, for the first time in our work, definitely above the lower world of life. There is nothing to show that the animals below man have any conception of the spiritual. It is true that there are various statements on record which seem to indicate in some animals, the horse and the dog, for instance, a dread of unseen powers, a recognition of some element in nature which is invisible to the eyes of man. But what these facts indicate, what influences affect the rudimentary intellect of these animals in such instances, no one is able to [226] say. Though some vague recognition of powers or existences beyond the visible may arise in their narrow minds, it does not probably pass beyond the level of instinct, and doubtless lies almost infinitely below man's conception of the spiritual. In this stage of intellectual development, then, we have to do with a condition which seems to belong solely to man, or has but a germinal existence in the lower organic kingdom.

In fact, primitive man may well have been as devoid of the conception of a realm of spirit as was his anthropoid ancestor. The lowest savages of to-day are almost, if not quite, lacking in such a conception, and are destitute of anything that can fairly be called religion. Where apparent religious ideas exist among them we cannot be sure to what extent they have been infused by civilized visitors, or how far ardent missionaries, in their anxiety to discover some trace of religion in savages, have themselves inadvertently suggested the beliefs which they triumphantly record. The Pygmies of Africa, the Negritos of Oceanica, and various debased tribes elsewhere,

157

may possibly be quite destitute of native religious conceptions, at least of a higher grade than those which move the horse and dog to a dread of the unseen. It should be borne in mind that these tribes have for thousands of years been in some degree of contact with more developed races and subject to educative influences, and the crude religious conceptions [227] which some travellers attribute to them may well have been derived, not original.

Investigation in this field certainly gives us abundant warrant to believe that primitive man, on whose mind no influences of education could act, was destitute of religion, and that man's conception of the unseen arose gradually, as one important phase of the development of his intellect. Any attempt to trace the stages of this religious development is far beyond our purpose, even if we were capable of doing it. It must suffice to say that man everywhere, when he emerges into history as a semicivilized being, is abundantly supplied with mythological and other religious conceptions which indicate a long preceding evolution in this field of thought.

For extended ages the realm of the unseen has been acting upon the mind of man; filling him with dread of malevolent and reverence for beneficent powers, inspiring him to acts of worship, peopling his imagined heavens with imagined deities, and giving rise to an extraordinary variety of deific tales and mythological ideas. The literature of this subject would fill a library in itself, and is almost abundant enough to supply one with reading for a lifetime. Yet it is largely, if not wholly, ideal; it is in great part based on false conceptions and misdirected imaginings; it rarely adduces evidence, and such evidence as is offered is always questionable; in short, scientific investi [228] gation and the critical pursuit of facts have taken no part in the development of religious systems, and a deep cloud of doubt envelops them all.

It is by no means our purpose to seek to throw discredit on any of the great religions of the world. To say that they have been products of evolution is not to invalidate them. Much that is true and solid has arisen through evolution. To say that they lack scientific evidence is not to question their validity. Many of the subjects with which they deal lie beyond the reach of scientific evidence. Science has hitherto dealt strictly with the physical; it has made almost no effort to test the claims of the spiritual. In fact, the highest of these claims, that of the existence of a deity, must lie forever beyond its reach. God may exist, and science grope for Him through eternity in vain. Finite facts can never gauge the infinite. Proofs and disproofs alike have been offered of the existence of an infinite deity, but the problem remains unsolved. None of these proofs or disproofs are positive; they all depend on ideal conceptions, and ideas are always open to question; positive facts on either side of the argument are, and are always likely to be, wanting, and the belief in God must be based on other than scientific grounds.

But when we come down to the lower levels of the domain of the spiritual we find ourselves on firmer ground. Here we are dealing with the finite, not with the infinite, and nothing that is [229] finite can lie beyond the boundaries of investigation, however long it may take to reach it. The question of the existence of spirits, for instance,—that much mooted problem of the immortality, or at least of the future existence, of man, which forms so prominent an element in modern religion,—dwells within the possible reach of science, and the attempt to deal with it by scientific evidence may reasonably be made. When we pass beyond the realm of the senses we find ourselves in a kingdom peopled by stupendous forms and forces,—space, time, matter, energy, and perhaps infinite consciousness,—all in their ultimate conditions too vast for the finite mind to grasp, all presenting problems open to speculation, but beyond the reach of demonstration. But below these lie finite possibilities which the human mind may now be, or may become, capable of comprehending, and prominent among these lies the problem just mentioned, that of the existence of a spiritual substratum in man, a soul which is capable of surviving the death of the body. This is a subject with which all of us are deeply and intimately concerned, and it may be well to close this volume with a brief glance at its status as a scientific question.

The belief in the immortality of man is comparatively modern in origin. There is no satisfactory evidence that any such belief existed among the old Jews, or that it arose in Palestine before the time of Christ. It arose at an earlier period in India [230] and Persia, but everywhere it was late in its appearance as a well-defined doctrine. Yet, while positive evidence is wanting, there can be little doubt that crude and vaguely formulated ideas of the existence of man after death have been very long entertained. The traditions of all peoples that have a faith above that of fetichism contain stories of the apparition of spirits of human origin, and when we reach civilized peoples and more advanced religions we find these in abundance. The annals of Christendom are full of them. They are equally abundant in the centres of other developed forms of faith. If we could accept these legends of the emergence of spirits through the thin veil that separates time from eternity as established facts, the problem would no longer need solution. As it stands, however, the great mass of such narratives are utterly lacking in evidence of a character which science can admit. They are bare, unsustained statements, thousands of which would be far outweighed by a single one fortified by demonstrated facts. Occasionally, indeed, the story of an apparition has been closely investigated, and there are a few cases of this kind handed down from the past which seem reasonably well established. But any statement coming from prescientific days is open to doubt; methods of investigation then were not what they are now; the dogma of the existence of spirit is too important a one to be accepted on any but incontroverti [231] ble evidence, and the vast sum of statements of apparitions which have come to us from the past, or from the non-scientific peoples of the present, must be dismissed with the one verdict, not proven.

There is one important fact, however, connected with the question of spiritual appearances, which is worthy of some consideration. It is a fixed rule in the history of opinions that beliefs founded on imagination or misconception have declined with the advance of enlightenment, and many conceptions, once strongly entertained, have faded and vanished in the light of new thought, or where retained have been so only by the ignorant and unreasoning. It is of interest to find that this has not been the case with the belief in spiritual manifestations. This has held its own to the present time, and, while it is largely sustained by the unintelligent and credulous, it can claim a considerable body of intelligent adherents to-day, even in the most enlightened nations. This belief, known as spiritism, with the manifestations upon which it is founded, lies open, therefore, to modern scientific investigation; and this has been, to some extent, applied to it, with, in various instances, rather startling results.

It is certainly of significance to find that a number of prominent scientists, thoroughly skilled in the arts of investigation, have attacked this problem with the purpose of annihilating it, and have ended in becoming convinced of the truth of spirit [232] ism. It may suffice to mention two of the most striking instances of this. In the early days of the spiritist propaganda, Robert Hare, a famous chemist of Philadelphia, entered upon an investigation of the so-called spiritual phenomena with the declared purpose of proving them to be fraudulent. His observations were long continued, his tests varied and delicate, and he ended by himself ardently adopting the belief he had set out to abolish. Somewhat later William Crookes of London, an equally famous chemist and physicist, entered upon a similar investigation, and with like results. The tests applied by these men were strictly scientific, and of the exhaustive character suggested by their long experience in chemical investigation; and their conversion to the tenets of spiritism, as a result of their experiments, was a marked triumph to the advocates of the doctrine. Various others of admitted high intelligence, who made a similar investigation and were similarly converted, might be named. Two of the best known of these were Judge Edmonds, of the circuit court of New York, and Alfred Russel Wallace of England, who shared with Darwin the honor of originating the theory of natural selection.

While these, and others of scientific education, were converted to spiritism, many investigators came to an opposite conclusion, while a similar negative result was reached in the investigations of several committees of scientists. The latest [233] and most persistent attempt to search into the reality of phenomena of this character has been that made by the London Society for Psychical Research, whose investigations have extended over years and have yielded numerous striking and suggestive results. The most important conclusion at which the members of this society have so far arrived is the hypothesis of Telepathy, or the seeming power of one mind to influence the thoughts of another, occasionally over long distances, in a method that appears analogous to that of wireless telegraphy. The evidences in favor of this doctrine are so numerous that it has been somewhat widely accepted, and the title applied to it has come into general use. It indicates, if true, remarkable powers in the mind of man, capabilities that seem far to transcend those of the ordinary intellectual activities.

This is one side of the case. The other side now calls for presentation. This is that the great body of scientists utterly reject the theory of spiritism, and look upon its manifestations as due to fraud, misconception, credulity, or some other of the weaknesses to which human nature is liable. As regards the opinions arrived at by the prominent scientists mentioned, these men are looked upon by their fellows of the great scientific body as mentally warped, or as having allowed themselves to be victimized by impostors. The fact that Professor Crookes has continued one of the [234] most acute and deep searching of investigators into the phenomena of physics, and that his results in this direction are accepted without question, and that Professor Wallace is acknowledged to be one of the leading thinkers of the day, has not sufficed to clear them of the doubt which rests upon their sanity or their critical judgment in this particular, and the very attempt of any one to investigate the so-called spiritual manifestations is widely looked upon as an evidence of credulity or some greater mental weakness.

This result may seem singular, yet it is not without abundant warrant. It must be borne in mind that the phenomena in question differ essentially in character from those with which science is usually concerned. The field of scientific investigation is distinctly the material; the facts with which it deals are those apparent to the senses, or which can be tested by material instruments; its discoveries are generally susceptible of but one interpretation; its methods are capable of being indefinitely repeated, and its results, if justly interpreted, are unvarying in character. None of these postulates fully applies to the spiritistic investigation. Here the conditions differ, the results vary, the methods can rarely be exactly repeated, conscious beings, instead of unconscious instruments, are the agents employed, and the secret thoughts and purposes of such agents are very likely to vitiate the result, and open a field of [235] doubt which does not exist in the investigation of the inorganic world.

This is one of the causes of the doubt of scientists. It is not the only or the chief cause. The latter is the fact that the claims of spiritism lift man into an entirely new domain of the universe, remove him from the great field of the material with which he is physically affiliated and to which his senses are closely adapted, and place him in a region beyond the scope of the senses, a vast kingdom which is held to underlie or subtend the physical, and which the ordinary outlook of the scientist fails to perceive. It requires no strain of the imagination to admit the existence of a new constituent of the atmosphere. It requires a great strain to admit the existence of a new constituent of the universe, a vast spiritual substratum to the domain of matter. Religion, with its ideal tests, has long maintained this to be a fact. Science, with its rigid material tests, sternly questions it, and demands that the existence of an inhabited spiritual realm shall be incontestably proved by scientific evidence before it can be accepted.

This demand is a reasonable one. The world is growing rapidly more scientific, and the old method of arriving at conclusions is daily losing strength. Beliefs based on ideal or imaginative postulates, once strong, are now weak. Faith founded on ancient authority is active still, but promises to become obsolete. The way of science is growing [236] to be the way of the world, and in the time to come intelligent men will doubtless demand incontestable evidence of any fact which they are asked to accept.

As regards the phenomena in question, however, it cannot be said that they have been fairly or fully investigated by scientists. They have been set down as the work of charlatans, and their apparent results ascribed to fraud, collusion, credulity, and mental obliquity in general. The fact, that of the scientists who have exhaustively investigated the spiritistic phenomena, a considerable number have accepted them as valid, has had no effect upon scientists as a body, who, in this particular, occupy the position which they accuse nonscientists of maintaining, that of forming opinions without investigating phenomena.

This attitude of the scientific world toward these problematical occurrences is quite comprehensible. Throughout the nineteenth century the attention of scientists has been almost wholly directed toward the investigation of the forms and forces of matter, the phenomena and principles of the visible universe. In this they entered, at the opening of the century, upon an almost virgin field, which they have wrought with great diligence and with remarkable results. It is very possible, however, that in the twentieth century no such undivided allegiance will be given to the phenomena of matter, but that the attention of scientists [237] will be largely diverted from the physical to the psychical field of investigation, which may prove to be a far broader and more intricate domain than we now have any conception of.

Psychical phenomena have attracted some attention during the recent century. One by one the problems of hypnotism, unconscious cerebration, double consciousness, telepathy, spiritism, and the like, all at first set down as unworthy of consideration, have forced themselves upon the attention of observers, and each of them has been found to present conditions amply worthy of investigation. This work has hitherto been performed by occasional individuals, but the number of workers in experimental psychics is steadily increasing, and their domain of research broadening, and we may reasonably look forward to results approaching, perhaps exceeding, in interest those reached in material investigation.

There is a whole world before us, that of the mind and its phenomena, fully equal in interest and importance to the world of matter, and presenting as numerous and difficult problems. Hitherto it has largely been dealt with from the ideal or metaphysical standpoint; only recently has it been subjected to physical analysis, and already with striking results. During the century before us it is likely to attract a wide and active circle of investigators, with what results it is impossible to predict. This is the only way in which [238] the problem of the existence or non-existence of a spiritual life can be solved to the satisfaction of those of a scientific turn of mind, and this solution must be left to the future to attain.

In the present work we are concerned with man's past rather than his future. It is what man has come from, not what he is going to, that forms the subject of our inquiries. We have been led into these remarks simply as an outcome of a brief consideration of man's relations to the spiritual element of the universe, and may close our work with the suggestion that the problem of human evolution may be immensely greater than that involved in the study of the ancestry of man.

THE DAWN OF REASON Or, Mental Traits in the Lower Animals

By JAMES WEIR, Jr., M.D.

Author of "The Psychical Correlation of Religious Emotion and Sexual Desire" etc.

16mo. Cloth. $1.25

Review of Reviews.

"This book presents evidences of mental action of the lower animals in a clear, simple, and brief form. The author has avoided technicalities, and has also resisted the temptation of the psychologist to indulge in metaphysics. Dr. Weir has relied for evidence on the results of his own independent study of biology at first hand, disregarding the second-hand data used by many of the authors once regarded as standard authorities in this department of research."

The Nation:

"The title raised in our mind some vague fears that we might find physiology and psychology mixed up inexpertly with metaphysics; but we see in the writer a close observer, who takes his stand on firm ground, and goes into the objective world of animals for his facts."

THE MACMILLAN COMPANY

66 FIFTH AVENUE, NEW YORK

www.ingramcontent.com/pod-product-compliance
Lightning Source LLC
Chambersburg PA
CBHW031631210526
45464CB00004B/1852